The Field Description of Metamorphic Rocks

The Geological Field Guide Series

Field Hydrogeology, Fourth edition Rick Brassington

Drawing Geological Structures Jörn H. Kruhl

Field Hydrogeology, Third edition Rick Brassington

Basic Geological Mapping, Fifth edition Richard J. Lisle, Peter Brabham, and John W. Barnes

The Field Description of Igneous Rocks, Second edition Dougal Jerram and Nick Petford

Sedimentary Rocks in the Field, Fourth edition Maurice E. Tucker

Field Geophysics, Fourth edition John Milsom and Asger Eriksen

Field Geophysics, Third edition John Milsom

The Field Description of Metamorphic Rocks

SECOND EDITION

Dougal Jerram

Centre for Earth Evolution & Dynamics, University of Oslo, Norway
DougalEARTH Ltd, Solihull, UK

Mark Caddick

Virginia Tech
Blacksburg, VA, USA

WILEY Blackwell

This edition first published 2022

Edition History
John Wiley & Sons Ltd (1e, 2013)

Registered Offices
John Wiley & Sons, Inc., 111 River Street, Hoboken, NJ 07030, USA
John Wiley & Sons Ltd, The Atrium, Southern Gate, Chichester, West Sussex, PO19 8SQ, UK

Editorial Office
9600 Garsington Road, Oxford, OX4 2DQ, UK

For details of our global editorial offices, customer services, and more information about Wiley products visit us at www.wiley.com.

Wiley also publishes its books in a variety of electronic formats and by print-on-demand. Some content that appears in standard print versions of this book may not be available in other formats.

Library of Congress Cataloging-in-Publication Data

Names: Jerram, Dougal, author.
Title: The field description of metamorphic rocks / Dougal Jerram, Centre for Earth Evolution & Dynamics, University of Oslo, Norway, DougalEARTH Ltd, Solihull, UK, Mark Caddick, Virginia Tech, Blacksburg, VA.
Description: Second edition. | Hoboken, NJ : Wiley, 2022. | Series: The geological field guide series | Revised edition of: The field description of metamorphic rocks / Norman Fry. 1984. | Includes bibliographical references and index.
Identifiers: LCCN 2021052480 (print) | LCCN 2021052481 (ebook) | ISBN 9781118618752 (paperback) | ISBN 9781118618691 (adobe pdf) | ISBN 9781118618677 (epub)
Subjects: LCSH: Metamorphic rocks–Handbooks, manuals, etc.
Classification: LCC QE475.A2 F79 2022 (print) | LCC QE475.A2 (ebook) | DDC 552/.4–dc23/eng/20211204
LC record available at https://lccn.loc.gov/2021052480
LC ebook record available at https://lccn.loc.gov/2021052481

Cover Design: Wiley
Cover Image: Courtesy of Susanne Schmid

Set in 8.5/10.5pt TimesLTStd by Straive, Pondicherry, India
Printed and bound by CPI Group (UK) Ltd, Croydon, CR0 4YY

C9781118618752_220222

CONTENTS

PREFACE – THE FIELD DESCRIPTION OF METAMORPHIC ROCKS

In many regards, metamorphic rocks represent some of the most complicated challenges that you will find in the field. They were obviously once igneous, sedimentary, or even different metamorphic rocks, and you may be able to interpret some of this early history at the outcrop. But they have subsequently been changed through combinations of pressure, temperature, and reaction with fluid so that they might now look radically different to their original form. In order to understand these changes (the rock's metamorphosis), and to some extent the original 'parent' rock, the field description of metamorphic rocks requires careful observation and a grasp of many aspects of the broader range of the geosciences. It is not enough just to know how to identify key metamorphic minerals in the field: multidisciplinary skills borrowed from other branches of field geology, and even engineering, are increasingly essential requirements for the modern metamorphic geologist.

This concise guide is designed to give students, professionals, and keen amateur geoscientists the key tools needed to help understand and interpret the origin and evolution of complex metamorphic systems in a focused way, while in the field. This extensively revised and reorganised colour guide builds on Norman Fry's original version, published in 1984 as part of the (then) Geological Society of London Field Guide Series. Since 1984, much has changed in the scientific community's understanding of metamorphic processes and in the ways that fieldwork is conducted. Accordingly, we have tried in this first colour revision to incorporate much of this newer thinking and methodology. At the same time, we have aimed to remain true to the original philosophy of a portable guide that concisely explains the basic concepts underpinning the field description of metamorphic rocks. Indeed, we have kept and built on some sections of Fry's original text. The original version was necessarily limited by its black and white printing, and we have enjoyed updating the figures in this version: almost every figure in the original book has been replaced here or reproduced in colour. We hope that the inclusion of the new colour images and a simple, colour-coded index system will help the reader to navigate their way through the different types, grades, and origins of metamorphic rocks. Both authors grew up with the original versions of the Geological Field Guide Series and one of us (Dougal) was even taught by Norman Fry at Cardiff University. So it has been a great pleasure, if not a long journey, to revise this field handbook, now published as part of Wiley Blackwell's 'Geological Field Guide Series'. We hope you find this new guide a great companion and an essential aid when confronted, perhaps for the first time, with metamorphic rocks in the field.

Dougal Jerram and Mark Caddick 2021

Preface – Meet the Authors

Dougal holds a 20% research professorship at the University of Oslo and is the director of DougalEARTH Ltd. He is primarily a field geologist and has undertaken fieldwork all over the world and experienced a wide range of Earth's geology and landscapes from Africa to Antarctica. He started his geological career in the UK, where he cut his teeth on the many fundamental outcrops the UK has to offer through a classic Geology degree at Cardiff and a PhD at Liverpool. His main expertise is in rock microstructure and textural analysis, 2D–3D modelling of rock textures, and understanding aspects of volcanic rifted margins from a hard rock basis. In recognition of his early significant contribution to Earth Sciences, he was awarded the Murchison Fund of the Geological Society in 2006. Dougal has written a number of other books centred around the Earth Sciences for both adults and children. He is also keen on science outreach and has been a presenter on Discovery channel's *The Very Edge of China* (2019), *Hardest Job* (2017), BBC's *Fierce Earth* series (2013–14), *Operation Grand Canyon* (2014), as well as appearances on National Geographic, Smithsonian Channel, Eden, Channel 4, and Abandoned Engineering.

Mark is an Associate Professor at Virginia Tech. A metamorphic petrologist, his work focuses on the micro-scale processes that lead to changes in rock mineralogy and texture, and the tectonic-scale processes that these may reveal. He has worked on metamorphic rocks from a variety of settings and with a wide range of styles, spanning from cold, deep subduction to high temperature crustal melting. Though much of his research is lab based or computational, it invariably starts in the field. He also works on high energy impacts and high temperature reactions of minerals in jet engines – which he obviously thinks of as a form of metamorphic geology! Mark was a student in the UK, at the universities of Bristol and then Cambridge, before moving to Switzerland as a research scientist at ETH Zurich. He has been in Virginia since 2012, during which time he has taught, amongst other things, an introduction to the geosciences, igneous and metamorphic petrology, thermodynamics, and field-based courses.

ACKNOWLEDGMENTS

First and foremost, we would like to thank Norman Fry, whose original book was an important guide for both authors as students, and he is also thanked for giving us the go ahead to update the book and to provide all the original materials that formed an invaluable framework as we planned this revised version. This book has taken a long time to mature and we must thank the support and patience of the team of people at Wiley Blackwell and associated editorial groups (both past and present) who have helped to get the book finished. We must particularly thank (and apologies if we miss anyone); Mandy Collison, Andrew Harrison, Frank Weinreich, Emma Cole, Shiji Sreejish, Priyadharshini Arumugam, Bobby Kilshaw, Athira Menon, Nithya Sechin, Vinodhini Mathiyalagan, Audrie Tan, Fiona Seymour, Ian Francis, Delia Sandford, and Rachael Ballard. Many people directly contributed figures and photographs and input to this book scientifically and these are particularly thanked for their help and open sharing of information, including (in no particular order); Isabela Carmo (with additional help from Prof. Renata S. Schimdt – UFRJ), Hans Jørgen, Nick Timms, Steve Reddy, Richard Brown, John Schumacher, John Howell, Susanne Schmid, Jim Talbot, Christoph Schrank, Bob Tracy, Claudio De Morisson Valeriano, Clayton Grove, Dave Prior, Tonje Lund, Nigel Woodcock, Victor Guevara, and Chris Clark. David Gust, Scott Bryan, Jess Trofimovs, and the Queensland University of Technology team are thanked for access to the QUT metamorphic teaching samples.

Dougal is particularly grateful to the people who showed me some of the classic metamorphic terrains, intrusive contacts, and regional geology, where I learned much about these systems. Wes Gibbons, Dave Prior, John Wheeler, Lee Mangan, Bob Hunter, Mike Cheadle, and Henry Emeleus introduced me at the early stages of my career to some of the classic Scottish locations, and more recently the likes of John Schumacher and Torgeir Andersen introduced me to some of the more exotic and incredible metamorphic textures I have seen. My colleagues in Oslo such as Trond Torsvik, Henrik Svensen, Sverre Planke, Olivier Galland, Bjørn Jamtveit, François Renard, Stephanie Werner, Karen Mair, Brit Lisa Skjelkvåle, Bernd Etzelmüller, Carmen Gaina, and the whole of the CEED team over the last 10 years have shown great support, particularly to my book writing efforts and research collaborations, and further afield my "brother" Breno Waichel is thanked for exposing me to the South Atlantic Margins and to many Brazilian colleagues. Jo Garland and Izzy Jerram are thanked for their ongoing support, and particularly Jo for proof reading and figure commenting at various stages. Finally, I would like to personally thank all those that have helped in discussions in the field all over the world where complex hard rock relationships have been made clearer by great collaborations (you are soooo many, and you know who you are, cheers!).

Mark would like to thank Alan, Mike, Nigel, Tim, Jon, and the other great mentors he has met along the way. He only knows about many of the outcrops photographed in this book thanks to the generosity of friends and colleagues such as Eric Reusser, John Schumacher, Filippo Schenker, and Bob Tracy. The current and former members of the Metamorphic Processes group at Virginia Tech, and the students of VT's GEOS 2024, 3704, and 4964, have always provided the best reasons to go back out and teach in the field, and I'm particularly grateful to those of you whose fingers, arms, and feet crept into some of the photos in this book – you know who you are! Thanks to Christiana Hoff for commenting on earlier versions of some chapters. Finally, thanks to my wife, Kristie, who read parts of the text, commented on many of the figures, and had the good grace to remain patient with me throughout this whole process.

INTRODUCTION AND OCCURRENCE

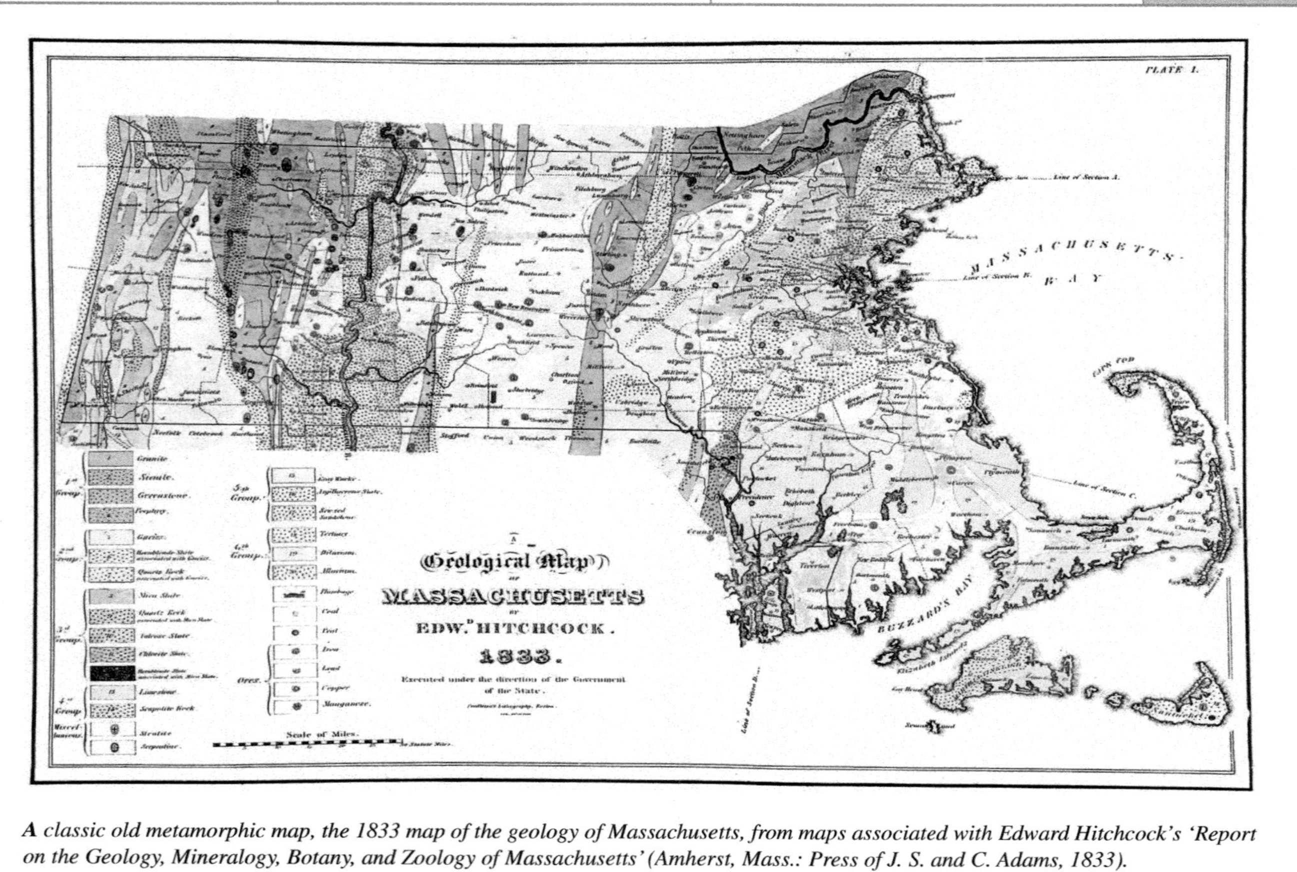

A classic old metamorphic map, the 1833 map of the geology of Massachusetts, from maps associated with Edward Hitchcock's 'Report on the Geology, Mineralogy, Botany, and Zoology of Massachusetts' (Amherst, Mass.: Press of J. S. and C. Adams, 1833).

1
INTRODUCTION AND OCCURRENCE

Metamorphic rocks form a substantial proportion of the material that makes up the Earth's crust, and metamorphic processes have been almost continually occurring throughout geological time since the origin of that crust. Metamorphism can be defined simply as the process by which sedimentary or igneous rocks are transformed (metamorphosed) by re-crystallisation due to changes in pressure, temperature, or fluid conditions. To complicate matters somewhat, metamorphism can of course also act on rocks that have already been metamorphosed previously, building layer upon layer of complexity into those rocks that record field evidence of some of Earth's most dynamic processes. Our understanding of metamorphism is somewhat limited by the fact that we are unable to directly observe it happening to the rocks. As you read this, metamorphism is in action all around the planet, in all aspects of the Earth's plate tectonic system (e.g. Figure 1.1), but we cannot directly see it (generally because it happens at depth and very slowly). In order to understand the processes and products of metamorphism and alteration in rocks, detailed fieldwork, petrography, experimental studies, and numerical modelling are required. It is important to note, however, that the very origin of metamorphic petrology (the science of understanding the distribution, structure, and origin of metamorphic rocks) is rooted in a tradition of careful and systematic field observation, and that this remains an absolute cornerstone of the discipline today. Since the late nineteenth century, Earth scientists have strived to develop an understanding of metamorphic processes by identifying the different types of key minerals, mineral assemblages, and structures present in the metamorphic rocks. Using these observations and knowledge of some fundamental principles, mineral reactions can be calculated and/or experimentally derived to help explain and understand the process by which the original rock was metamorphosed into its current state. These rocks often encode evolving conditions at tectonic plate boundaries, so deciphering their mineralogical history may be thought of as a window into the crustal-scale processes that form, modify, and stabilise Earth's crust. Underpinning all of this is the petrologist's ability to identify, describe, relate, and collect metamorphic rocks in the field, and it is these skills which this book aims to explore and impart, by its use in the field description of metamorphic rocks.

1.1 The Importance of Fieldwork in Metamorphic Terrains

In many ways, metamorphic geology requires you to be skilful in most aspects of the Earth sciences. As metamorphic rocks can be formed from any original rock (the parent rock henceforth being called the protolith), an ability to identify and be familiar with the wide variety of minerals and textures of sedimentary and igneous rocks is a general requirement for any budding metamorphic geologist. Additionally, as the very processes involved in metamorphism are commonly associated with deformation, a keen understanding of structural geology and tectonics is also needed. *In many ways, the metamorphic scientist needs to be a jack of all trades and a master of one!*

Due to the potential complexity within metamorphic rocks, the importance of careful fieldwork cannot be overstated. The different types of observation that can be made at various scales in metamorphic terrains allow the student/researcher to build up a list of clues, like in a forensic study,

The Field Description of Metamorphic Rocks, Second Edition. Dougal Jerram and Mark Caddick.

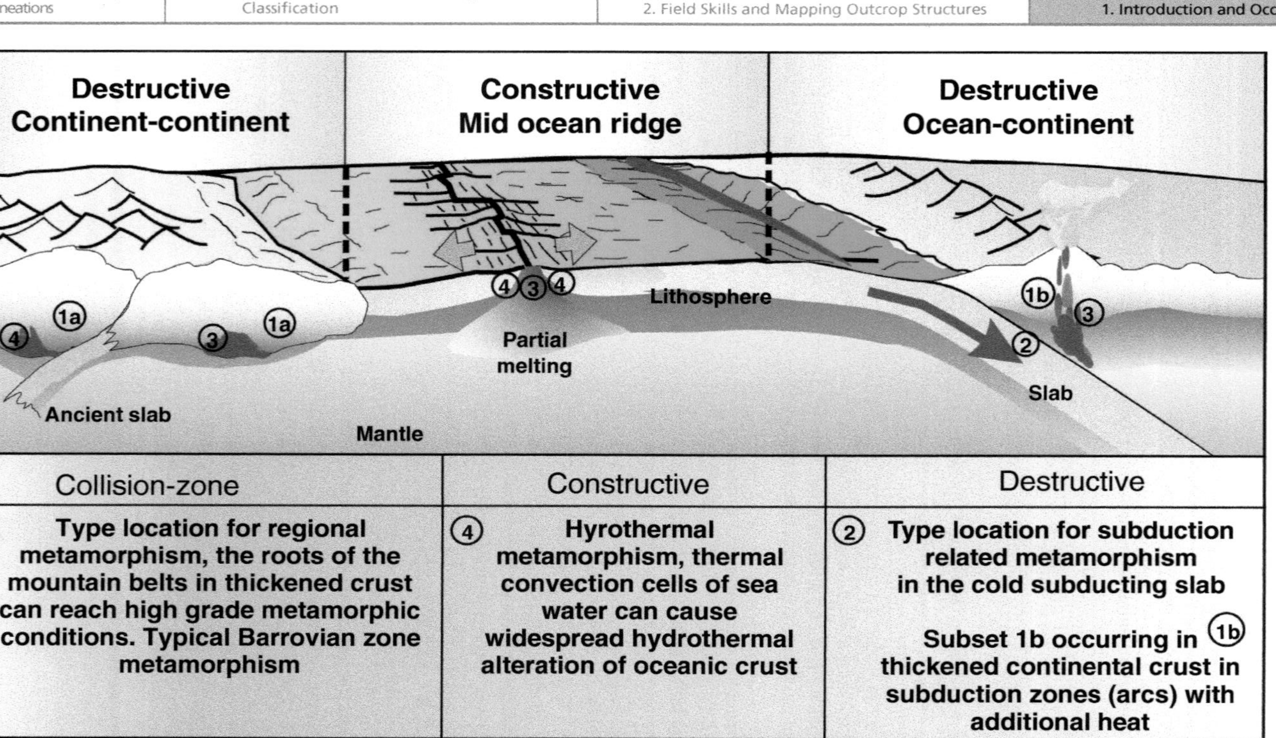

Figure 1.1 *Schematic of the plate tectonic settings where metamorphism is occurring around the world (see also Figure 1.2).*

which can be used to help derive the type of metamorphic rock, its protolith, and the range of processes that it has undergone to reach its present state. The map-scale distribution of metamorphic rocks can reveal the processes that formed them, but as we discuss in the following chapters, the correct interpretation of even the smallest parts of a field area are rooted in good field observations. This book aims to help build you skills in this area! Careful identification of rocks and structures is all the more important when taking samples from the field back to the laboratory for further study and analysis. The record of structures within and around the rock mass may ultimately help you to better interpret features you subsequently see down the microscope or the data that you receive from laboratory analysis.

Describable features which can be observed in metamorphic rock masses include:

1. *Pre-metamorphic* – e.g. bedding and other sedimentary features, contact relationships between batches of melt, or even fossils (though in most cases the features may be altered beyond normal recognition).
2. *Metamorphic* – relating to local mineral changes due primarily to changing temperature and pressure.
3. *Metasomatic* – involving the chemical transport and mineral change associated with fluids.
4. *Structural* – relating to and recording the rock's deformation at any point in its history.

Limitations exist as to how much information one can record regarding any of these features without the need for microscopic and chemical measurements, which is the realm of specialist study that will be touched upon within this book but is not our major theme. With good field observations of mineralogy, texture, and structure, one should still be able to adequately describe the rock masses in terms of their types and occurrence, hopefully also being able to build up an inference of the evolving conditions of their formation. Such description is particularly appropriate for the production of geological maps, logs, and recordings of outcrop structures, which will be covered in more detail in Chapter 2.

This book forms a companion to the other texts in the geological field guide series, e.g. *The Field Description of Igneous Rocks*, *Sedimentary Rocks in the Field*, and *The Mapping of Geological Structures*, and as such does not cover in detail the pre-metamorphic features of sediments and igneous bodies that may sometimes be preserved in metamorphic rocks. We do, however, show many examples of these in cases where they can either be shown to help in the identification of the protolith rock or reveal something fundamental about the metamorphism itself (e.g. that it happened in the presence or absence of deformation). There is substantial overlap between the skills required to be a metamorphic geologist in the field and those considered to be the realm of a structural geology, at least in terms of fieldwork measurements/observations, and particularly when mapping in metamorphic terrains. As such, this text will aim to provide as much help in terms of structural description, as will be necessary to get the most out of your metamorphic rocks. The reader will need to make an assessment as to what level of understanding of sedimentary, igneous, and structural geology might be best suited for the problem at hand, and where needed can supplement this guide with an appropriate partner guide. For example, if you are mapping a metamorphically altered igneous region, then additional help from *The Field Description of Igneous Rocks* may be useful. In a thrust zone, the structural guide may provide some vital additional assistance, and so on. However, we have tried, wherever possible, for this book to be a stand-alone guide to achieve success in the field description of metamorphic rocks. Ultimately, we aim for this handbook to provide the required information on how to observe metamorphic rocks in the field, from the outcrop to the hand specimen scale, and to tie these observations into basic interpretations of how the metamorphic rocks formed. This also necessitates comments on sampling strategies for projects in which fieldwork is the start of a wide-reaching study. As such, before we take on metamorphic rocks in the field it is useful to consider how metamorphism relates to regional and global tectonics and the main occurrence of metamorphic rocks.

1.2 Understanding Metamorphism; Pressure/Temperature Relationships

Rocks undergo metamorphic and metasomatic changes as they are subjected to different pressure and temperature conditions, or are infiltrated by chemically reactive fluids. Indeed, a fundamental building block to a deeper understanding of metamorphism is a good grasp of pressure, temperature, and time (it takes time for metamorphic reactions to take place, evidence of which may be preserved in the field in the form of incomplete reactions). In this sense, it is very useful from the onset of your training as a metamorphic Earth scientist to become familiar with the ranges of pressure and temperature experienced in the Earth and the key metamorphic mineral associations (assemblages) that are found within these ranges. One of the main ways in which we consider this is through what is known as a *P*/*T* diagram, in which changing aspects of a rock are plotted as a function of pressure (*P*) and temperature (*T*). This allows one to highlight various aspects of metamorphism and question how they might be represented in the field. *P*/*T* diagrams will appear throughout this text to help understand the types and styles of metamorphism, and will feature specifically in Chapter 3 in relation to the main classification of metamorphic rocks, and in associated tables within the reference Chapter 8.

At this introductory stage it is useful to consider the basic *P*/*T* diagram in relation to the relative intensity of metamorphism, as this forms a good basis for understanding under what conditions the different types of metamorphic rocks are formed. Figure 1.2 shows a *P*/*T* diagram (with approximate depths included) that expands on the key 'facies' concept (originally described by Pentti Eskola in 1915), namely that rocks of a similar composition will, when subjected to the same *P*/*T* conditions, form the same mineral assemblages. You can also see how this relates to the main tectonic settings by referring the numbers on the trends to the locations on Figure 1.1. The fields in Figure 1.2 thus map out the *P*/*T* stabilities of major mineral assemblages that could form in a metamorphosed mafic rock (e.g. a basalt) as a general reference. A far more detailed and subtle record of mineral reactions almost certainly occurs in most rocks and will be discussed in subsequent chapters, but the reactions at the boundaries of these fields are significant enough that the metamorphic facies (and thus approximate metamorphic *P*/*T* conditions) of a mafic rock can generally *be identified in the field*. Generally speaking, Figure 1.2 suggests that low grade metamorphism starts around 150–200 °C and ~3 kbar (300 MPa, or ~10 km depth). As temperature and pressure increase, the grade of metamorphism progressively increases accordingly until, at temperatures of 600–800 °C (or greater), the rocks themselves begin to melt and we start to enter the realm of igneous petrogenesis. These fields and the main ways in which we classify metamorphic rocks will be discussed in detail in Chapter 3, and as you go along you will see that the *P*/*T* of the rocks can be displayed in a variety of diagrammatic forms.

1.3 Mode of Occurrence of Metamorphic Bodies

Because metamorphism is a response of pre-existing rocks to changes in temperature and pressure, it may be expected that metamorphism is restricted to major zones of deformation in the Earth, such as convergent (destructive) tectonic plate margins. Clearly where major tectonic forces act, such as at subduction/collision zones, the crust undergoes deformation, and rocks will experience changing pressure and temperature upon burial as the crust is thickened. However, metamorphism is not restricted to these environments of the Earth. Extreme temperature changes can be achieved through the contact of molten igneous bodies (sills, dykes, magmas chambers) with country rocks. Also, in certain settings, the wholesale circulation of fluids through the crust can lead to alteration and metamorphism (such as at mid-ocean ridges). Rocks that are metamorphosed in subduction/collision zones undergo metamorphic changes over broad zones, and can record evidence of passage from one metamorphic grade to another as they journey through different depths. These form the most common types of metamorphic rocks, termed the Regional Metamorphic Rocks. Where rocks are metamorphosed due to contact with hot igneous bodies they are referred to as Contact Metamorphic Rocks.

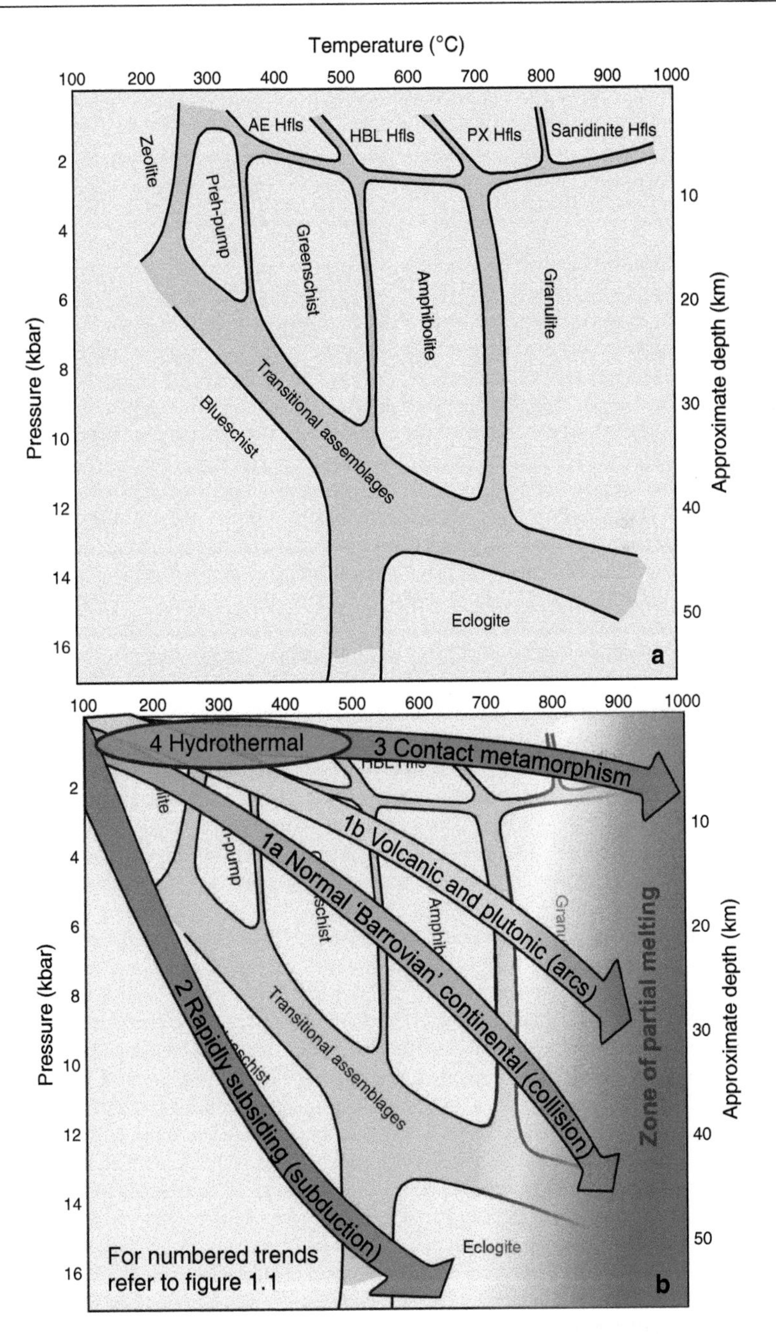

Figure 1.2 *The P/T diagram: (a) the classic fields of metamorphism of mafic rocks (the so-called matamorphic facies) in P/T space, and (b) the routes that certain tectonic systems take through the P/T space which give rise to different metamorphic rocks. This will be expanded on in more detail in Chapter 3.*

Finally, where alteration and metamorphism occur due to fluids, the rocks are called Hydrothermally Altered Metamorphic Rocks. Some more exotic and rare examples of metamorphic rocks include those specific to fault zones (Cataclastic Metamorphism) occurring as a result of mechanical deformation when two bodies of rock move past one another, and Shock Metamorphism (Impact Metamorphism), where rocks are metamorphosed due to impact from an extraterrestrial body, such as a meteorite or comet.

1.3.1 Regional metamorphic rocks

The most common of the metamorphic styles, regional metamorphism, occurs in zones defined by key pressure and temperature environments found at certain burial depths in the Earth's crust. The regional metamorphic zones are also restricted by certain tectonic settings that are generally related to subduction and continental collision zones, defining two broad groups of regional metamorphic rocks: those related to mountain building events, where two continents collide, and those formed at subduction zone settings where oceanic crust is subducted. As continent–continent collision is preceded by subduction, both styles of regional metamorphism can sometimes be found in the same location, occasionally with strong evidence of high-pressure, low-temperature mineral growth in a subduction zone overprinted by higher temperature mineral growth upon and after collision.

Regional metamorphic zones typically occur due to thickening and/or burial of the crust, so pressure is a very important parameter that drives reactions to progressively change the original rock (protolith) into its different metamorphic types. Certain reactions are strongly pressure-dependent, and thus the occurrence of specific minerals or mineral assemblages is indicative of ranges of pressure conditions (the most well-known of which is that diamond typically only forms at pressures greater than the base of normal continental crust). The pressure at which metamorphism occurred is often linked directly with depth, by considering the force applied by the overlying mass of rock. Temperature generally increases with pressure, known as the geothermal gradient (the rate at which temperature increases with depth), but the specifics of this gradient can vary dramatically with tectonic setting. Again, there is generally a mineralogical response to changing temperature, so unravelling the metamorphic history of a series of outcrops in the field can yield important information about the style of regional metamorphism and, thus, tectonic setting and evolution.

A classic study by G. Barrow in the late nineteenth and early twentieth century examined a suite of metamorphic rocks from the Scottish Highlands. The rocks here were formed as part of a mountain building event, the Caledonian orogeny, and show progressively increasing grades of metamorphism depicting the different depths of burial and temperatures attained during the mountain building event. This was unknown at the time of Barrow, whose work was subsequently expanded on by people such as C.E. Tilley in the 1920s, and the concept of exposure of the roots of an ancient mountain belt was yet to be developed. Barrow, however, recognised that specific metamorphic minerals occur in certain groupings, and that the order of their occurrence was predictable (if he walked north in one valley and first found rocks containing garnet, then found rocks containing staurolite, he would be able to find the same succession in a parallel valley several kilometres away). These minerals, termed 'index minerals', were thus used to define different zones of metamorphism in the Scottish highlands, and have since been termed the Barrovian sequence. Based on a concept called isograds (planes of constant metamorphic grade), in which the first appearance of a key metamorphic index mineral is mapped, six 'Barrovian' zones were thus defined (Figure 1.3). These will be touched on in more detail in Chapter 3, but can be considered as the background to many of the main metamorphic rocks recognised in unroofed collision zones. As such, a systematic view of metamorphism as formulated by Barrow would state that if the protolith was an aluminous sedimentary rock (e.g. a shale) a typical sequence from low to high grade would exhibit the indicator minerals (which are the mapped isograds in Figure 1.3):

Chlorite > Biotite > Garnet > Staurolite > Kyanite > Sillimante > Melt

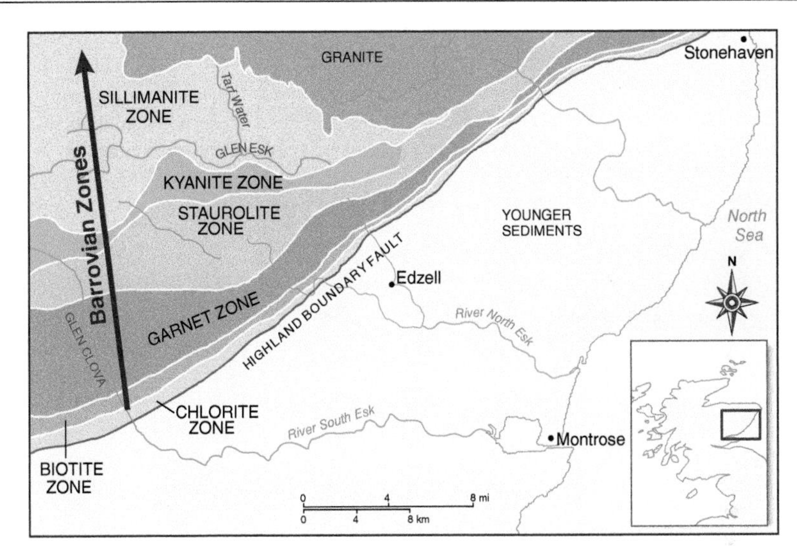

Figure 1.3 *The classic Barrovian Zones of regional metamorphism first described from Scotland.*

This mineralogical change with increasing grade would be mirrored in a maturation of the rock texture as follows:

Sedimentary Rock > Shale > Slate > Phyllite > Schist > Gneiss > Migmatite

If the starting material was an igneous rock, such as basalt, the sequence would be:

Basalt > Greenschist > Ambibolite > Granulite

This is highlighted by the 'normal continental (collision)' arrow in Figure 1.2. Examples of low intermediate and high grade regional metamorphic rocks are given in Figure 1.4. Chapter 3 provides a more detailed and systematic overview of how rock texture and mineralogy change in rocks of various compositions as metamorphic grade increases.

1.3.2 Subduction zone rocks

In a subduction zone setting (see Figure 1.1), the relatively fast burial of one of the cold plates leaves insufficient time for it to heat up substantially until it is at significant depth. It takes time for the subducted rocks to heat up (generally by conduction from the hotter rocks around them at depth), but application of pressure during burial is instantaneous. Thus in subduction zones the conditions of high pressure – low/moderate temperature metamorphism occur (e.g. the numbered trend 2 in Figure 1.2), and rocks exhumed from these settings record evidence for having been in the blueschist facies. Again, these rocks are characterised by certain indicator minerals and mineral assemblages, and blueschist rocks often appear blue (hence their name) because of the prevalence of a blue amphibole mineral called glaucophane. Exposures of blueschist facies rocks are relatively rare, most obviously because it is difficult to exhume them from the subduction zone to the Earth's surface, but they are an important record of plate tectonics on Earth and will be described in more detail in Chapter 3. An example of a blueschist is given in Figure 1.5a. When the most extreme pressures and moderate

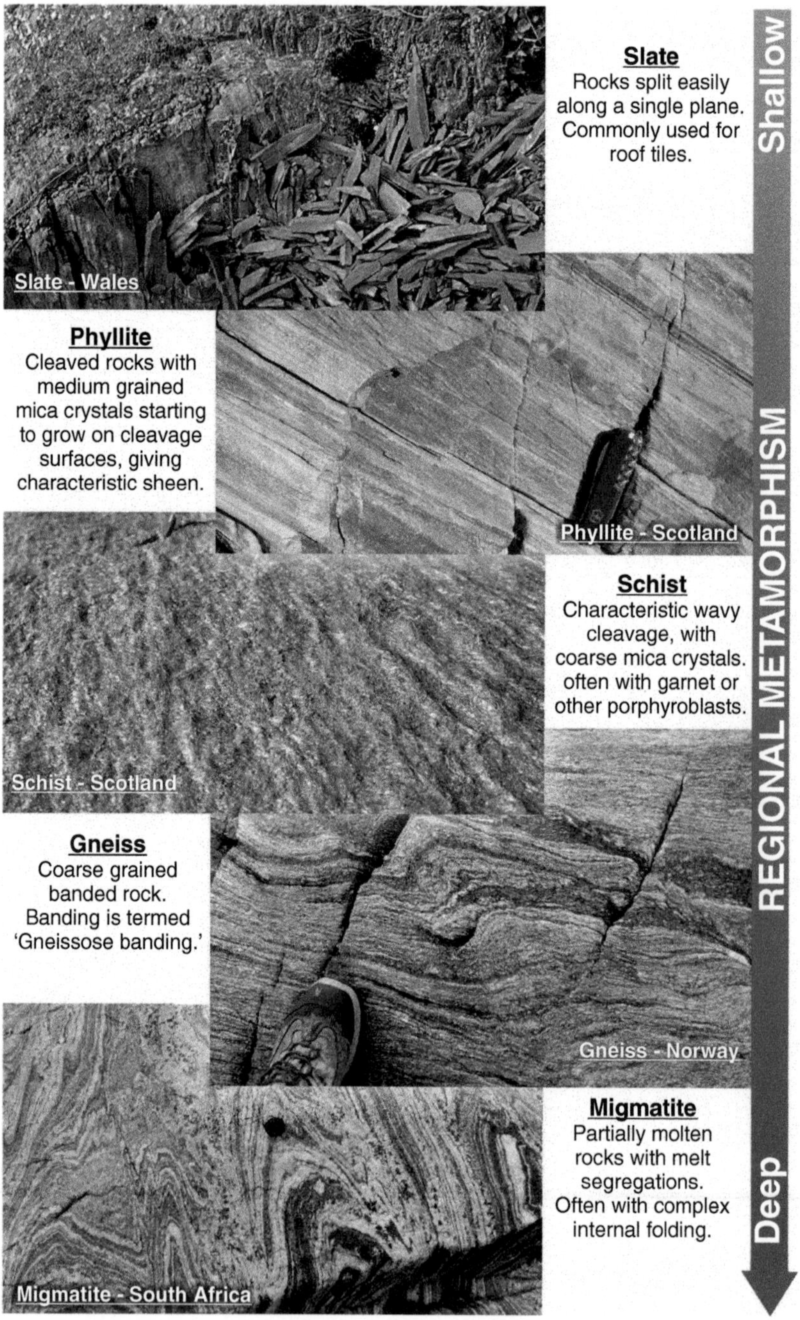

Shallow

Slate
Rocks split easily along a single plane. Commonly used for roof tiles.

Slate - Wales

Phyllite
Cleaved rocks with medium grained mica crystals starting to grow on cleavage surfaces, giving characteristic sheen.

Phyllite - Scotland

Schist
Characteristic wavy cleavage, with coarse mica crystals. often with garnet or other porphyroblasts.

Schist - Scotland

Gneiss
Coarse grained banded rock. Banding is termed 'Gneissose banding.'

Gneiss - Norway

Migmatite
Partially molten rocks with melt segregations. Often with complex internal folding.

Migmatite - South Africa

REGIONAL METAMORPHISM

Deep

Figure 1.4 *Examples of classic (Barrovian) regional metamorphic rocks (slate photo Jim Talbot, phillite, schist, and gneiss photos Dougal Jerram, migmatite photo Mark Caddick).*

Figure 1.5 (a) Blueschist facies, Syros, Greece (Mark Caddick for scale) with inset figure highlighting lawsonite porphyroblasts, (b) Eclogite facies, Alps (photo a Mark Caddick, photo b Hans Jørgen).

to high temperatures are reached, a group of rocks termed the eclogite facies form. Exposure of these on Earth's surface is again relatively rare, but they are generally easily identified, characterised by a pale green pyroxene (sodic rich called omphacite) and a deep red garnet (almandine-pyrope), an example of which is given in Figure 1.5b (see Chapter 3 for more detail).

The regionally metamorphosed rocks are often also characterised by having many structures associated with deformation. The rocks are put under pressure from all sides, but often this pressure is not the same from all sides. This leads to asymmetry in the pressure distribution and the alignment of new metamorphic minerals, rotation of existing and newly growing ones, and faulting and folding of the rocks during their metamorphism. These textures will be touched on in detail in Chapters 4 and 5, but banding, cleavage, folding, and dislocation structures are commonplace in regional metamorphic areas (e.g. Figure 1.6).

1.3.3 Contact metamorphic rocks

Igneous rocks can be emplaced into the crust at exceedingly high temperatures. Granites will crystallise at around 700+ °C, and basic rocks such as gabbro may intrude around 1200 °C, establishing a marked temperature gradient between the molten rocks and the host into which they intrude (commonly termed the 'country rocks'). Along the contact zones between the igneous bodies and their host rock, metamorphic reactions are commonly driven by heat from the cooling magma. This leads to a group of rocks called the contact metamorphic rocks. The contact or 'baked' zone around the igneous body can contain a variety of different metamorphic grades that are typically only seen over a relatively short distance as the effects of the hot igneous body diminish rapidly with distance from the magma. This zone of contact metamorphism is called the 'aureole' and is typically meters to tens of meters in thickness. Pressure tends to have little effect in contact metamorphism, as it is the act of

Figure 1.6 *Highly folded metamorphic carbonate turbidites, Namibia (photo Dougal Jerram).*

emplacing the hot igneous body and not a change in burial that makes the metamorphic aureole. Fluid flow during the metamorphism can substantially modify the wall rock composition, a process called metasomatism that is described more in Section 1.3.4, and can increase the footprint of the metamorphic effects by carrying heat further from the magmatic source (a process known as advection).

As with regional metamorphic rocks, different assemblages of minerals occur depending on the grade (mainly defined by the amount of heat) that the country rock reached, and depending on the type of country rock. With silliciclastic sediments like sandstones and shales the sequence may consist of chlorite, andalusite, and corderite hornfels, with silimanite and K-feldspar at very high temperature, and garnet if the crust was at sufficient depth during intrusion (e.g. a contact metamorphic overprint in a regional metamorphic setting). In limestone host rocks, marble is commonly formed, with tremolite, diopside, wollastonite, and forsterite as common minerals if the original carbonate was 'impure' (e.g. contained some Si). A schematic contact aureole with some examples is given in Figure 1.7 (further detail can be found in Chapter 6).

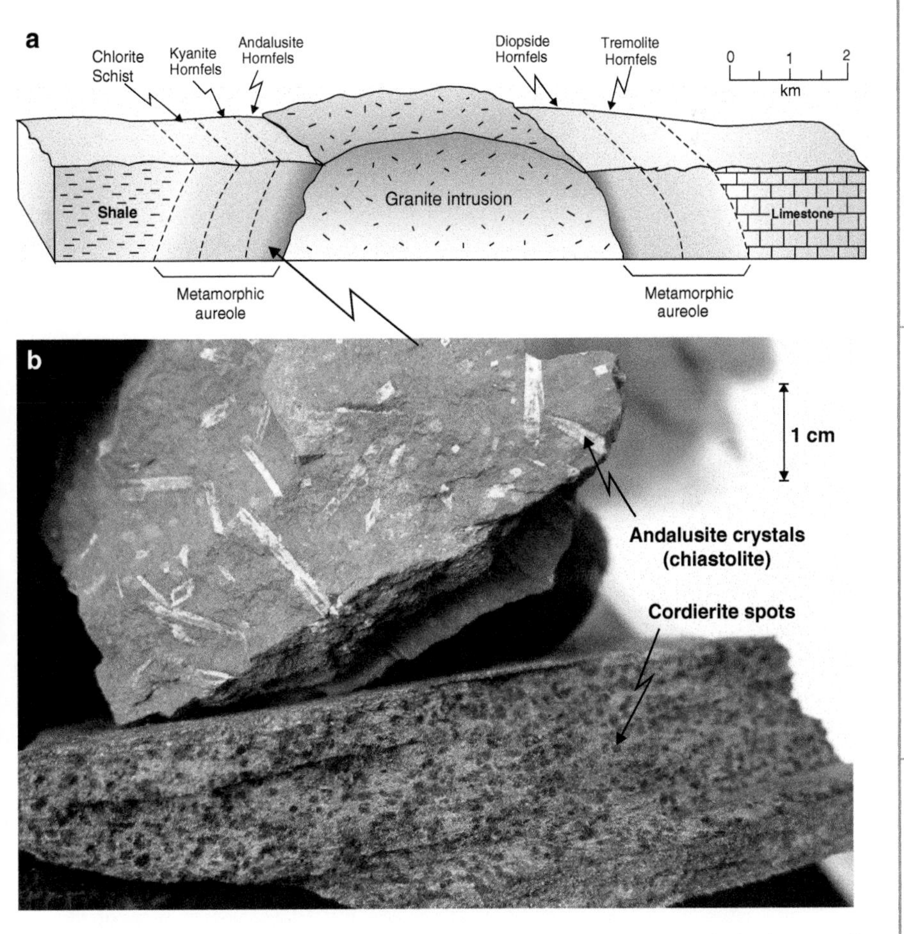

Figure 1.7 Contact metamorphism. (a) schematic of contacts around a granite body. (b) Examples of Andalusite (chiastolite form with graphite intergrowths) and cordierite hornfels from the Lake District, UK (photo Dougal Jerram).

A major difference between contact metamorphic rocks and the regional metamorphism discussed here is that contact metamorphism is generally quite static, with far less deformation during mineral growth. This means that the newly formed minerals are not typically as strongly aligned as they are in regional metamorphism, and an irregular orientation of fine grained minerals is typical of a 'hornfels', a classically diagnostic rock of relatively high temperatures of contact metamorphism.

1.3.4 Hydrothermal metamorphic rocks

The third major set of metamorphic rocks are formed through hydrothermal circulation of fluids. In hydrothermal alteration/metamorphism, the host rocks can be involved in wholesale chemical changes upon interaction with chemically reactive fluids. There exist large parts of the Earth's oceanic crust and upper mantle that are almost entirely made of hydrothermally metamorphosed rocks, formed as oceanic rocks interacts with large hydrothermal cells that circulate seawater and initiate mineral reaction. Depending on the temperature of the crust during this circulation and on the availability of water, the result can vary from almost pristine basalt with a little carbonate veining, through to highly 'serpentinised' rocks in which primary olivine in a mantle rock is thoroughly replaced by serpentine group minerals (see Figure 1.8).

The metamorphic rocks exhibit considerable chemical changes that are often termed 'metasomatic', with the loss of calcium and silica, and the relative gain of magnesium and sodium. In modern settings the occurrence of 'black smokers' and 'white smokers' on the sea floor are direct evidence of the hydrothermal cells in action. In the rock record, examples of obducted oceanic crust in the form of ophiolites display this hydrothermal metamorphism and they can also be associated with rich economic metal sulphide mineralisation.

***Figure 1.8** Serpentinised ocean crust from the Troodos Ophiolite, Cyprus. A fibrous serpentine vein is visible in the centre of the photo (photo Dougal Jerram).*

1.4 Summary

As we have introduced, a number of key settings exist that lead to the wide variety of metamorphic rocks and metamorphic associations that we find in the field (summarised in Figure 1.9; see also Figure 1.1). This field descriptions book is organised and laid out such that it introduces metamorphic rocks in terms of how they may be recorded in the field, making comments about sample collection and how work might be followed up in the lab and through additional studies (e.g. under the microscope). It is important to understand the various classification schemes and the different

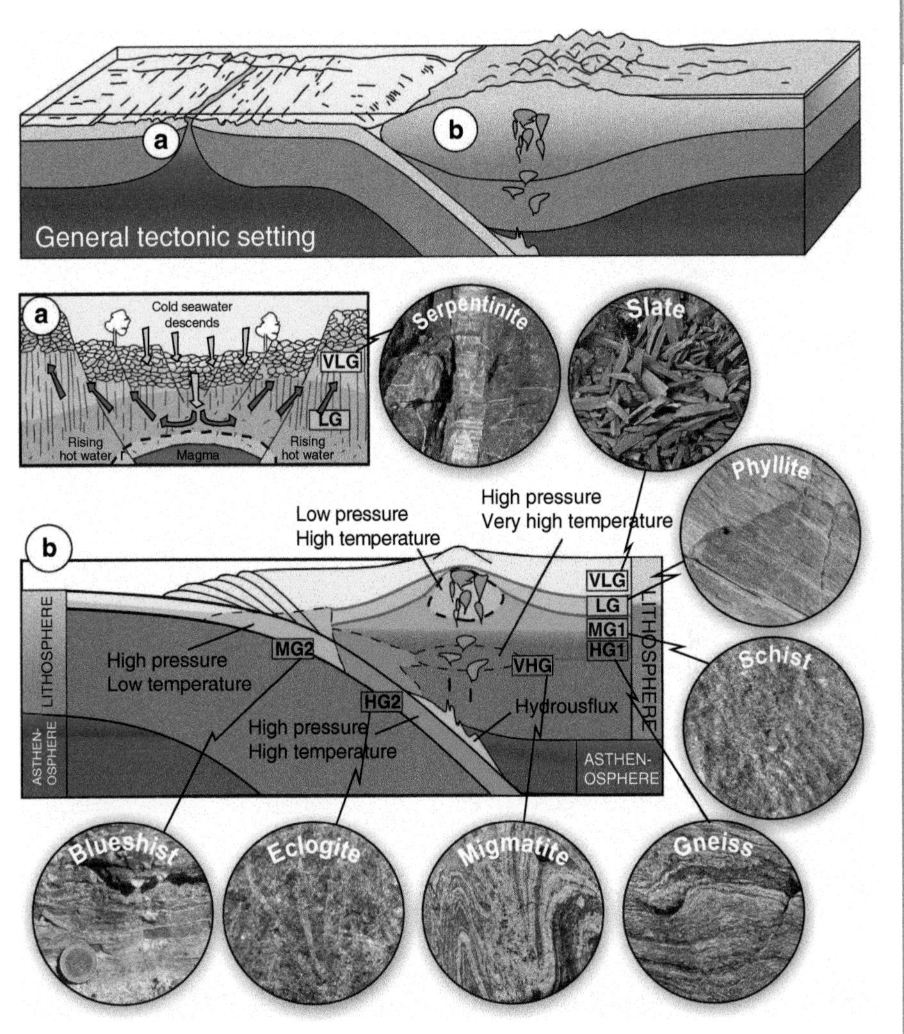

Figure 1.9 *Summary of the two main types of tectonic settings with an outline of the metamorphic grades (Source: partly redrawn from Hefferan and O'Brien (2010)). The metamorphic grades: Very Low Grade VLG, Low Grade LG, Medium Grade MG/MG2, High Grade HG/HG2 and Very High Grade VHG, are indicated on panels (a) and (b) (also refer to Table 3.1 and Chapters 3 and 8) (Source: Rock picture inserts from the figures in this chapter).*

grades of metamorphism that can affect the rocks, but it is also equally important to be aware of the many difficulties often faced in identifying metamorphic rock textures and assemblages. Chapters 2 and 3 provide advice and information on some of the key field skills and metamorphic outcrop description (Chapter 2), as well the main classification criteria (Chapter 3). Chapters 4 and 5 focus in more detail at key structures in metamorphic rocks such as cleavage, shistosity, and isolated bodies such as boudains, augen, and sheared entities. Chapter 6 looks at contacts with igneous rocks, veins, and reaction zones. Chapter 7 looks at the structural control of faults and fault zones in metamorphic rocks and shear zones including shear sense indicators. In Chapter 8, there are key tables and details for reference by way of a summary, as well as some advice for those who are undertaking a more detailed mapping dissertation or project.

FIELD SKILLS AND MAPPING OUTCROP STRUCTURES

Figure 2.1 *Be prepared! Picture of field kit; hard hat, compass clinometer, tape measure, mapping pens, maps, high-vis, jacket and so on. See the text for a detailed list.*

2

FIELD SKILLS AND MAPPING OUTCROP STRUCTURES

As with any form of geology, the key to good metamorphic study is to make meaningful and detailed observations at the gross and outcrop scale as well as at the finer scale. The best way to approach fieldwork is from a top-down approach where you assess the outcrop and field context as a whole and then focus in on the detail. Often, a detailed observation could have a number of possible interpretations, which are narrowed down considerably by placing it in the bigger context of the outcrop or region. A good metamorphic field geologist needs to have a good grasp of structural geology and also a basic understanding of igneous and sedimentary rocks (particularly their structures and textures), as you will probably need to measure the structural features associated with deformation of metamorphic rocks and interpret what the original 'protolith' rock was before metamorphism. The control that the protolith rock composition has on metamorphic mineral development is crucial, and will be outlined further in Chapter 3. In this chapter, we will introduce some of the basic skills and preparation that are required to get the best out of characterising metamorphic rocks at outcrop scale. We have deliberately included a number of real examples from field notebooks, etc., so that you can see what is realistically achievable and useful, and to get some indicators of different styles and approaches.

2.1 Equipment

Fieldwork studies require good equipment and planning in order to be best prepared for any eventuality and to provide the basis to undertake the best possible work. As with any outdoor activity, there is a likelihood of significant weather variation so a variety of clothing/footwear appropriate for the terrain and environment is required. Please take note of the geological fieldwork code, which contains detailed information about field safety, sampling, and general rules in the outdoors (see the British Geologists' Association for good documentation about this, e.g. https://geologistsassociation.org.uk/codesofconduct, and note that you can also refer to *Basic Geological Mapping* by Barnes and Lilse, from this geological field guide series). It would be good to have the fieldwork code with you to reference as required. In the modern age it is tempting to rely on smartphones, tablets, GPS, etc., for information about location; by all means, use these resources where you have them but make sure that you have the very basic field equipment at hand to complete your work. Batteries will run out, you may lose your device, and mobile service is still not always available in many field areas. It is more likely that someone will return or hand in a field notebook and pens, rather than a mislaid expensive device. Do not forget that *your safety* may be compromised if you do not have both a good set of maps and a good understanding of how to work with them and read from them. Additionally, careful background preparation before embarking on fieldwork will help you to be best prepared to deal with the variety of rock outcrops, required scales of observation, and variety of geology that you will encounter. Both basic and detailed geological fieldwork will often need to be done in a single visit to the outcrop, so the kit list for the prepared field scientist should include the following (see Figure 2.1 inset page):

The Field Description of Metamorphic Rocks, Second Edition. Dougal Jerram and Mark Caddick.

1. Introduction and Occurrence
2. Field Skills and Mapping Outcrop Structures
3. Metamorphic Minerals, Rock Types, and Classification
4. Understanding Textures and Fabrics 1: Banding, Cleavage, Schistosity, and Lineations

- General and detailed maps of the area of interest as paper copies as well as digital versions if available. This can also include, where available, aerial photographs, satellite data, and even geophysical maps, which can help with regional interpretations. Printed versions of maps available online (e.g. regional or national geologic surveys, or satellite maps, which are now readily available through NASA World Wind/GoogleEarth), can be very useful to help find your way around your field area and to locate good outcrops.
- Sturdy and preferably waterproof notebooks (take plenty to complete your task and *make sure* that you prefill each book with your name and contact details in case you mislay one), graph paper (different types and scales), a stereonet (with drawing pin/thumb tack), and tracing paper are useful in case you need to make detailed base maps and wish to plot structural data. You may also set up your own logging templates to record field data (see Section 2.3).
- A compass clinometer (check that it is set correctly for your position on the planet – information about magnetic declination can be found on most maps) and possibly a GPS hand receiver or GPS-bearing smartphone (again this must be checked for the correct datum and grid referencing system). It is vitally important that you record the settings of both your compass clinometer and your GPS into every notebook so that data can be corrected or converted later if need be. *Most* importantly, record every GPS point into your notebook as well, just in case you mislay your GPS. This provides a good back-up of your data and takes seconds in the field (do this at each locality, not back at base in the evening!!). Don't forget that while a GPS can only tell you your location, a good map, a compass, and good field skills can help you to accurately locate yourself *and* can help you to make crucial measurements of the attitude and nature of outcrops. A good smartphone app might also be able to do this, but ask yourself whether you would really want to be using your phone if it is pouring rain or whether you would rely on your phone's battery as your only way of locating yourself in the wilderness (not the safest of plans).
- A hand lens (a magnifying glass can also be useful), pocket knife, fine pencils for initial mapping, fine mapping pens (various colours), coloured pencils, rulers, clinometer, calculator, and grain size and fieldwork tables chart (a variety of small cards with basic info on grain size, sorting, and other key geological charts are available – and you can always construct your own). As you undertake more and more fieldwork, you are likely to develop your own style for recording observations, but it is clear that some people record data better with pencil, other people with pen. Either way, you eventually need a *neat* and *permanent* record of your observations, so make sure that you have a good selection of writing implements with you – over time you will naturally begin to gravitate to whatever works best for you. Some people also like to implement colour into their sketches, and this can be very powerful (see notebook examples shown throughout this and other chapters).
- Geological hammer, tampered chisels (the hammer needs to be appropriate for hammering rocks and metal, and the chisel also needs to be appropriate for the purpose – people have lost an eye using the wrong tools), and a variety of measuring tapes. You may also need sample bags, which are generally sturdy plastic or canvas bags that you can close securely and that can be labelled.
- Your primary method of recording information should be in your notebook, but a photographic record of outcrops is often useful, so take a camera or phone with a good camera. All photographs also need a scale, so make sure that you have a good scale bar with you (be it a person, a coin, a lens cap, or a hammer).
- For safety, you should have a hard hat and high visibility jacket/vest if you are going to be working anywhere near a road, near cliff faces, or in quarries. In fact as a general rule, high-vis clothing is quite useful in any location, should problems arise. Protective glasses and gloves are essential if you (or anybody in your party) will be hammering outcrops or samples (good gardening gloves can be of use when hammering and handling field samples). A first aid kit is essential, though this can be coordinated with other people in your party. A whistle and other survival equipment are advised if you are working in remote areas. In very extreme cases, you might need advanced survival and communication devices.

- If you are working in a restricted-access area (e.g. some private land) or anticipate hammering or sampling in a region where this is normally not permitted (e.g. a national park), make sure that you are carrying copies of whatever permissions or permits you have had to obtain to work there.

We have tried to make this list as detailed as possible to cover most eventualities. In many cases you would not have all of this equipment on your person at all times, but you will need to have easy access as required, e.g. at a base camp or field centre, or in a vehicle that you are using to access outcrops. It is also extremely important to understand that having the right equipment is absolutely pointless if you do not know how to operate it correctly – make sure you know everything important about your field equipment *before* you embark on each field season.

2.2 Preparing Maps and Basic Mapping

A detailed description of geological mapping is beyond the scope of this field description of metamorphic rocks, and one should refer to the specific mapping guide in this series. However, it is vitally important that the locality and general relationships of the metamorphic rocks that you are studying are recorded on appropriate maps: this also includes mapping out the distribution of the rock units at a variety of scales, and understanding how to refer key units between maps, logs, and notebooks. When mapping out metamorphic terrains, 1 : 10 000 is a commonly used metric scale (1 : 24 000 is more often used in the United States), but you may also be required to map and record information at different scales to capture the variations that are present in outcrop (see Section 2.3.3). You may even be mapping with overlays on satellite data where terrain maps are lacking, for example in remote areas. Base maps can be of varying qualities and more recently may have a number of colours on them. From our experience, we suggest that you create GREEN base maps from the topographic maps you have because having contours, roads, and other basic map ornaments in green can help emphasise the geology you map onto them and reduce confusion with any black or other colour ink symbols and annotations that you make. Green base maps can be achieved digitally or through photocopying/printing in just green toner.

If the size of the study area and the nature of the terrain allow, walk (or drive or ride) over the area, trying to view as much of it as possible from higher ground. You can use this as an opportunity to better understand the region with respect to your larger maps, aerial photographs, satellite maps, etc., to get a sense of scale, and to identify any particular areas in which outcrop exposure might be particularly good or poor. Take note of where it may be difficult to move around due to topography or vegetation, the main physical barriers (e.g. rivers or cliffs), and the main routes of access to different parts of the area. In this context it is good practice to construct a generalised base map of the region you are covering, highlighting where there is good access/outcrop, good mobile coverage, and key places of interest. This map can be duplicated so that you have a copy with you and at least one left back at base as a safety precaution, which can be referred to should people need to locate your likely position in an emergency. Again, it is easy to be complacent and think that your phone will be the key to safe navigation and location in your area; it is *not*.

Your generalised base map can include the approximate boundaries of the rock mass to be worked on. Note the approximate layout of outcrops, particularly if they are obviously of different rock types. Note for each one the extent and form of exposure (stream cuts, hilltop crags, riverbeds, roadside exposure), and whether exposures display obviously useful information (e.g. important minerals, textures, structures, or veins). At this stage you will begin to develop a general feel for the type of metamorphic terrain that you are looking at (e.g. regional, high grade contact metamorphism, exhumed subducted material, etc.; see Chapter 1), so it is useful at the reconnaissance stage to make a note of some of the key minerals, structures, and associations that you may expect in such terrains. Any obvious outcrop colour, jointing, weathering style, and overlying vegetation should also be considered at this point as this may help you to become more efficient in the field. Be aware that while these may correlate with rock type, they can also vary due to dependence on topography, drainage patterns, and other factors. Also note any minerals or geometric relationships that are likely to be troublesome to characterise and try to obtain additional information to help you with them. Remember when planning your field programme that you may need extra time for study, and you may need to find more specific information from books, papers, or existing maps to help you in areas not covered in detail in this handbook.

2.3 Notebooks and Data Recording

When fully prepared, you can start to undertake your primary task of recording observations about metamorphic rocks. At this stage, it may be tempting to rush up to an outcrop in order to identify rock types and minerals immediately and make what might perceived as rapid progress in the form of entries into your notebook. Be careful though; a more methodical approach is generally more productive when you first arrive at any new outcrop.

It is often useful to initially stand back and perform a reconnaissance overview of each outcrop. Spend a few minutes walking around the outcrop, observing its variety and possible key features. This is a far better use of time than to try and immediately take vigorous notes of details at the rock face. Remember to 'recce' your outcrops each and every time, as this allows you to gather your thoughts and prepare to record the information in a more organized way. It also gives you the highest chance of eventually focusing in on those parts of the outcrop that will yield the most useful and important information.

You should develop a clear structure to your notebooks. As already noted, it is good practice for each individual to develop their own style, but some things *must* be included in the style you adopt. It is vitally important to separate observations from interpretation, as well as to record information as accurately and quantitatively as possible. A poorly written notebook may blur observation and interpretation to such an extent that it becomes useless to the unobserved reader (or yourself if you read it many years, or even weeks, later!). Remember that the notebook should allow you to reconstruct your observations at some point in the future, potentially allowing you to come to different interpretations. Some people distinguish observation from interpretation by using different coloured pens, or by using different parts of the page for each – the method does not really matter, but the fact that there *is* a method does. Obviously, always record the date and your location (using a detailed grid reference or appropriate GPS position, but always clarifying how to read back the number at a later date). Additional observations such as the weather and your current mood seem frivolous, but are often useful pointers at a later date as you try to reconstruct aspects of the fieldwork. Don't be afraid to write detailed statements of your objectives and daily plans in your notebooks, as these can be very useful to help reconstruct what you were doing and how the thought processes has developed when you go back to the notebook once returning from the field (e.g. Figure 2.2). Use tables and figures as much as possible to organise data and to highlight outcrops, but also to record how your thought process is developing. You will need to define the key rock types (e.g. formations and members), identify key markers and contacts, and ultimately develop an understanding of the big picture of metamorphism and its development in your area. Some additional advice and guidance for geological description, defining formations and markers, developing the bigger picture, and final reporting is provided in Chapter 8 (Section 8.3).

2.3.1 Field sketches

Good interpretations require the right suite of key observations – don't leap into an interpretation too early. To help this you can start building a suite of observations and opinions while making a general outcrop sketch to which detailed observations can subsequently be added

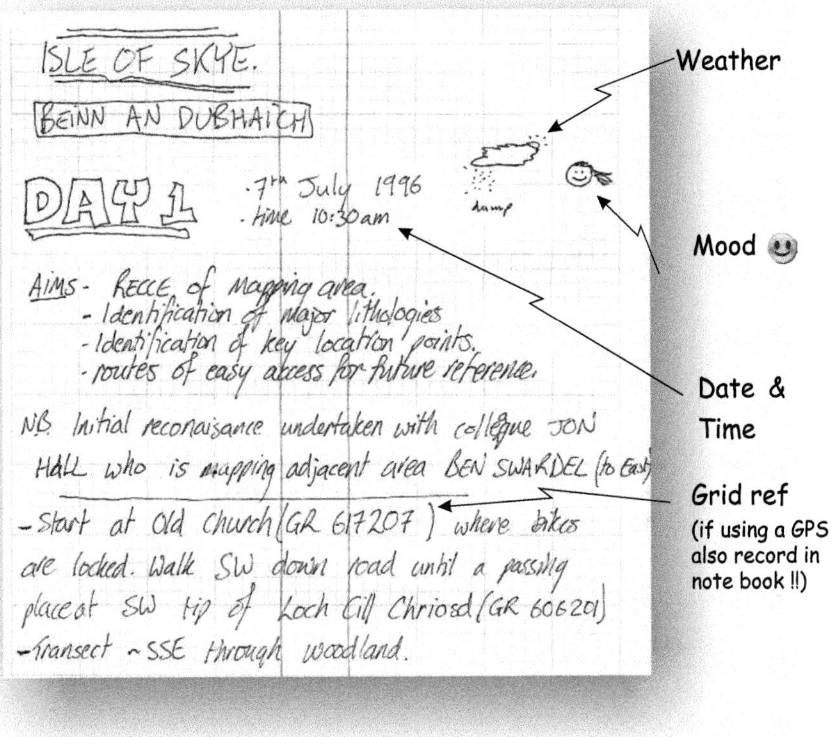

Figure 2.2 *Example of a simple but effective start of day layout. (example courtesy of Nick Timms).*

(e.g. Figure 2.3). Every sketch *must* have a scale and orientation (direction or field of view), and should be annotated as extensively as possible. Another good way to help build on an outcrop sketch is to have detailed sketch areas expanded from within the outcrop sketch to highlight particular features (e.g. Figure 2.4). Field sketches of key outcrops and features are particularly useful because they help you to make important observations and allow you to start discerning the sequence of events that may have led to the rocks that you are observing. It is good practice to also record in your notebook any samples and photos taken at the locality in question. As you develop your field notebook, sometimes small texture sketches can be useful to simply focus on particular detail, which in turn can also serve to help with building up your observations (e.g. Figure 2.5).

Due to the complex 3D nature of many outcrops, to discontinuous exposure, and to the relationships you are trying to get across, it may be best to construct one or several synoptic sketches, as a guide to how you think the rocks and structures have formed (e.g. Figure 2.6). These 'cartoon' like sketches can be very helpful to summarise the salient features of your outcrop and can be used to help explain processes, key relationships, and time sequences pertinent to your outcrop. Additionally sketched geological/outcrop maps can also help put everything into context (see Section 2.3.3).

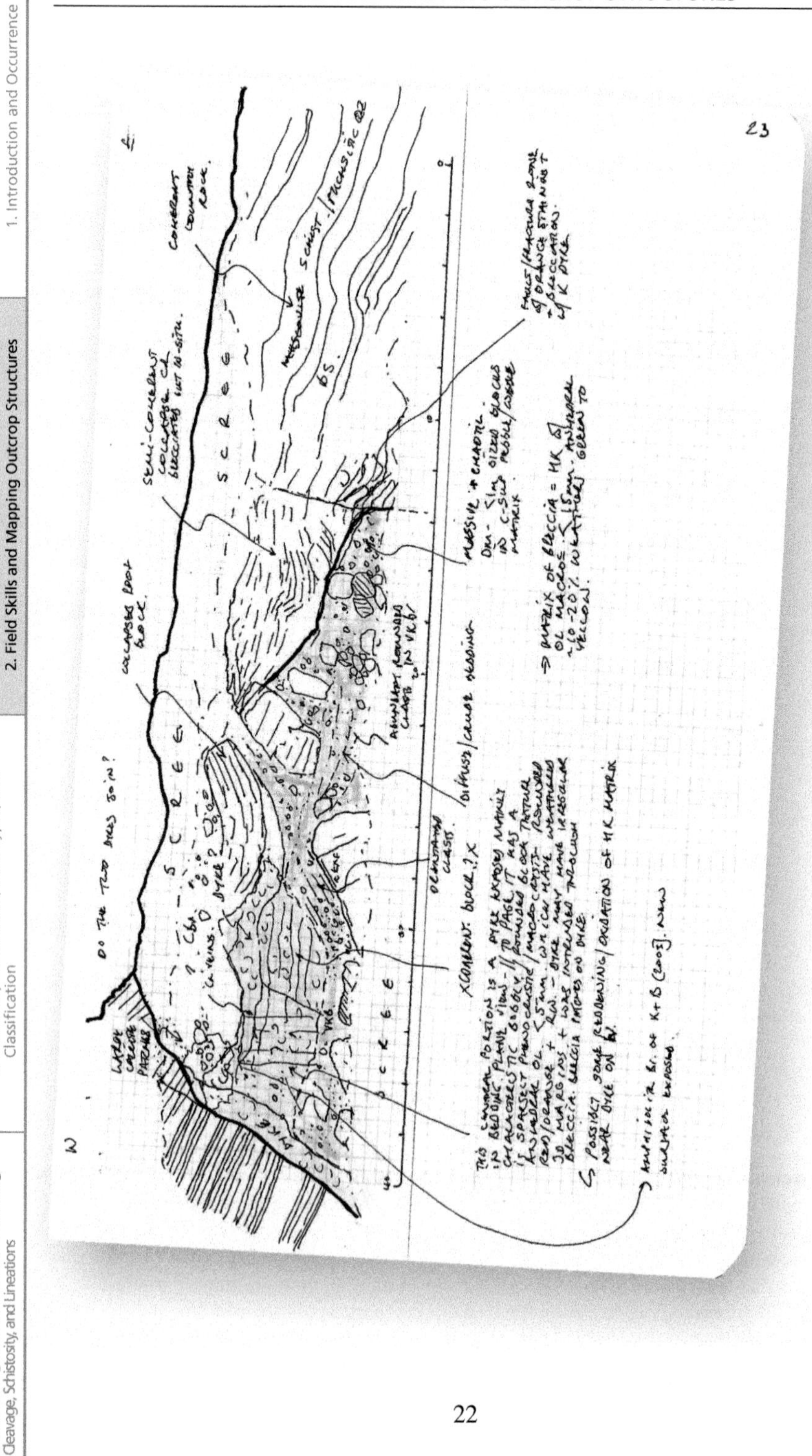
23
DO THE TWO DYKES JOIN?
COLLAPSED ROOF BLOCK
COHERENT COUNTRY ROCK
SCREE
DYKE

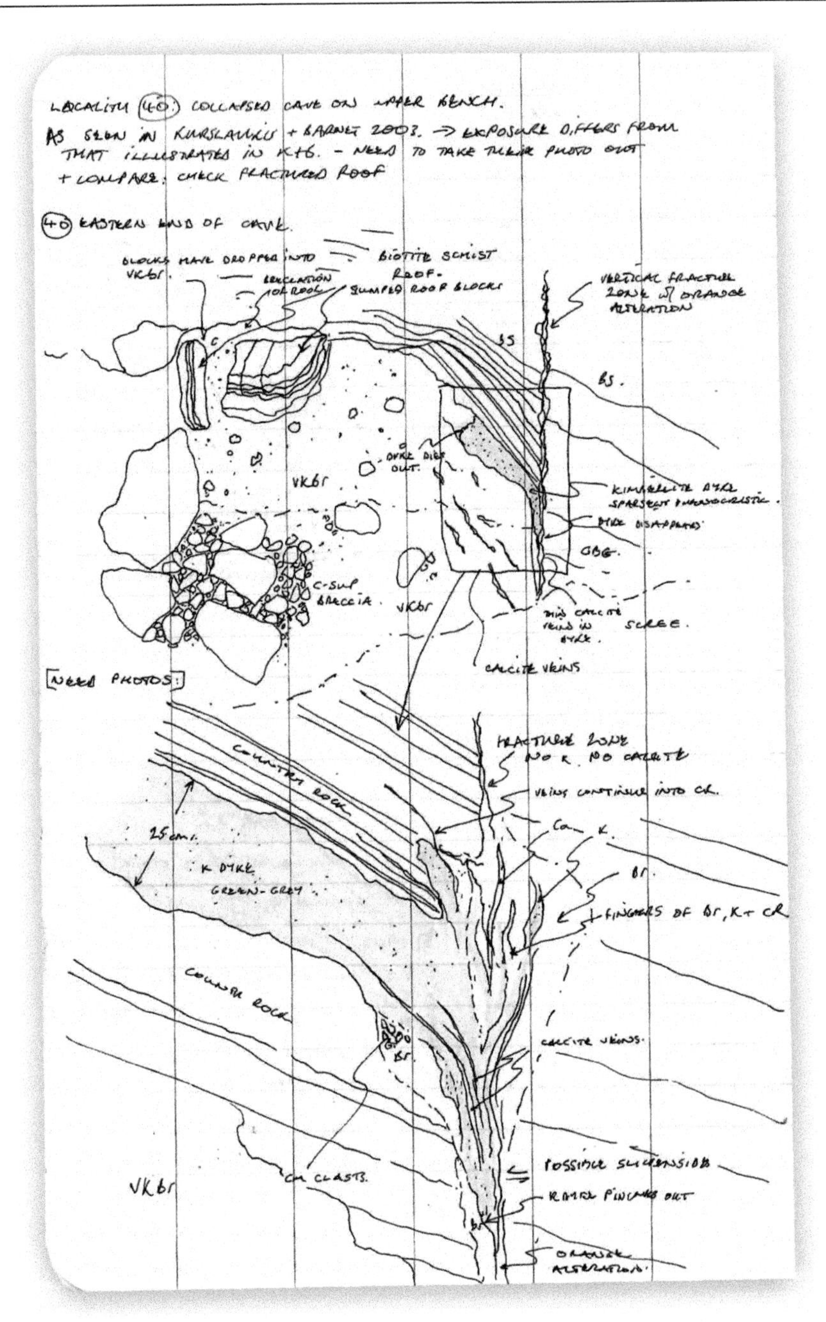

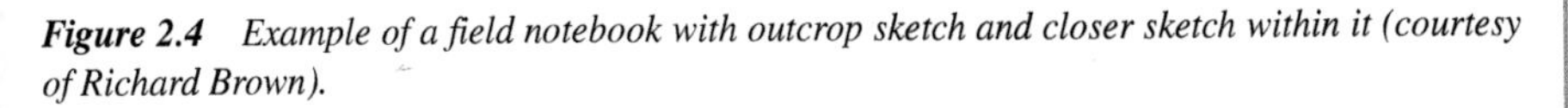
Figure 2.4 *Example of a field notebook with outcrop sketch and closer sketch within it (courtesy of Richard Brown).*

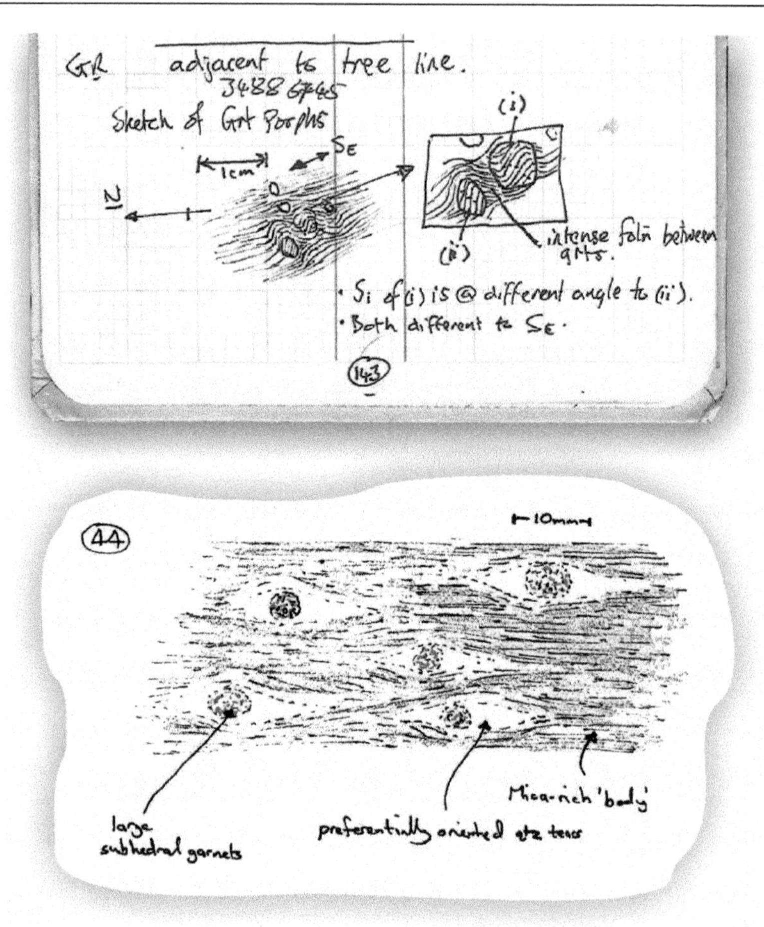

Figure 2.5 Examples of small, detailed, texture sketches (top courtesy of Nick Timms, bottom example Mark Caddick).

Remember, do not neglect descriptions of individual rock types. Without descriptions of all the rocks being correlated, correlation is ultimately a waste of time. Remember, also, that bedded, layered, or banded sequences may contain information on the nature of the banding (for example, cyclic variation in the sediments that were metamorphosed may help you to discern their origin, or faulted repetition may lead to an interpretation of tectonic process). This may not be apparent initially and may require time making additional observations elsewhere, or possibly constructing cross-sections or logs.

2.3.2 Logging – graphic logs

A graphic log can be a powerful way of portraying field data in metamorphic rocks, particularly where clear banding and layers occur. A log can be set up in a similar way to how you may log sedimentary or volcanic sequences, but can be modified to best represent the rocks in front of you. This

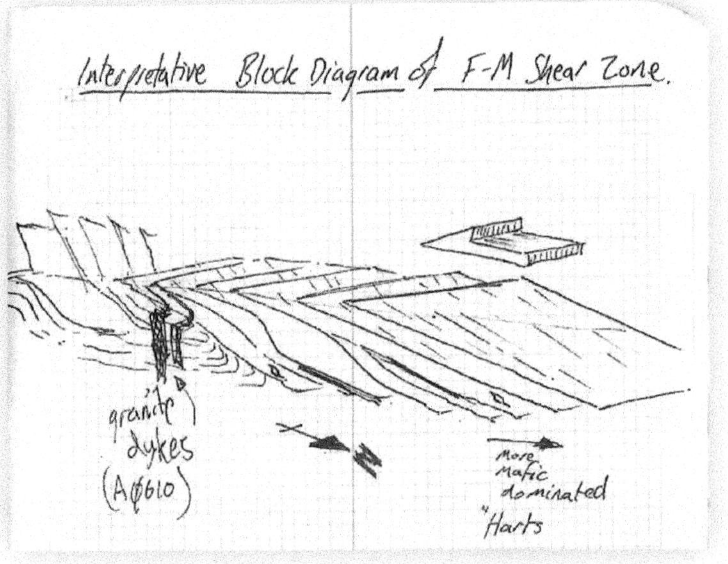

Figure 2.6 *Example of a field notebook with synoptic sketch (courtesy of Nick Timms).*

needs to be flexible, so you may set out a logging template to help you rather than conforming to a specified convention. An example is presented in Figure 2.7. In this case the first column is graphic. It includes thumbnail sketch caricatures of small structures, and can also have symbols marking particular mineral occurrences. The second column is a continuous trace representing the principal variable character (darkness in this example). The third column is for brief written notes. The log illustrated in Figure 2.7 was constructed to elucidate grading, so a fourth column has been added to show the graded units identified.

To help you see how graphic logs can be of use in metamorphic terrains, here are four cases where logs have proved useful:

1. Meta-sedimentary schists were logged in adjacent outcrops, in an attempt to establish a common lithostratigraphic succession, and for correlation between areas.
2. Meta-gabbroic schists were logged to portray cyclical compositional variations and grading at a scale of tens of metres in an attempt to distinguish original cyclicity from tectonic repetitions within the sequence.
3. A deformed greenschist sequence, seen to exist in places of previous sub-parallel sheet intrusions, was logged to discover whether there was symmetry from side to side of the sheets, as might be expected from a sub-vertical 'sheeted dyke complex', or whether a consistent asymmetry existed which might be interpreted as indicating a 'way up' of a set of sub-horizontal sills or lava flows.
4. Gneisses consisting mainly of alternating sheets of amphibolite and granite were logged to display graphically the manner in which the granite sheets became gradationally both more abundant and larger towards an area of granite gneisses, and to search for any accompanying changes in other rock-types and in tectonic structures.

Although symbols and abbreviations are used on logs as a way to portray information in a succinct form, a graphic log is of greatest help for those continuously variable rock characters

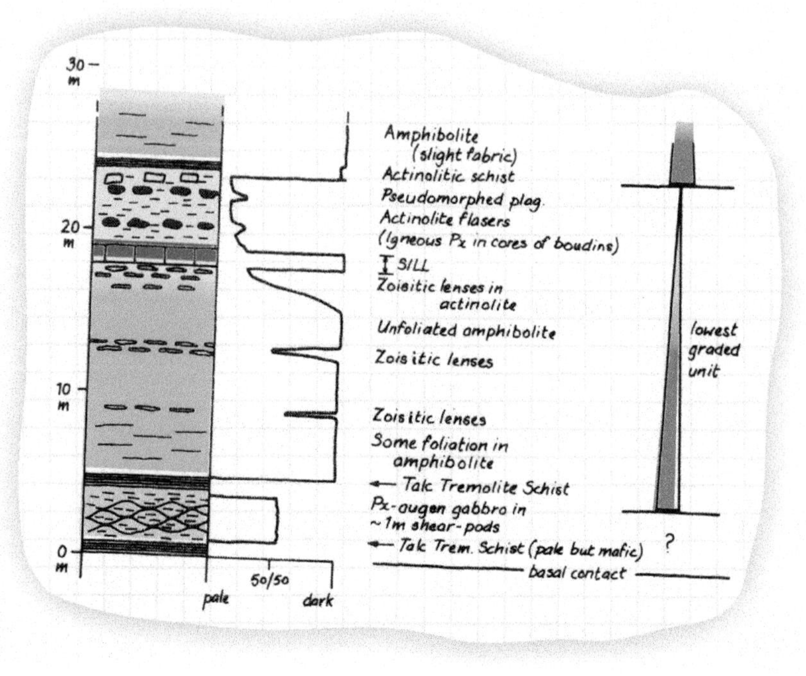

Figure 2.7 *Example of a graphic log highlighting a banded meta-gabbro sequence showing the modal abundance of light and dark minerals through the sequence.*

that may be represented by a column of variable width. In sedimentary logs, the clast size is the chief variable used to mark changes up the logged section. Grain size may be a useful character in some metamorphic rocks, whether of pre-metamorphic grains (clasts or igneous minerals), or of metamorphic minerals. In other cases, another variable, such as specific indicator minerals that can be easily measured or observed, may be a more useful primary characteristic. The key characteristic for the rocks in the examples listed above were (i) carbonate mineral content, (ii) colour (darkness), (iii) metamorphic grain size, (iv) whether granitic or not. In the first three of these (the fourth representing a simple yes/no case), chips of rock that are representative of end-members and intermediate states of the variable concerned were carried in the field for comparison during log construction. Collection of these reference samples is particularly important when trying to compare subtle variations in characteristics such as colour or texture. Figure 2.8 provides a detailed example of a log prepared for a set of subduction related strata in Syros, Greece.

2.3.3 What to record and mapping at different scales

Orientations of rock bands, veins, faults and contacts, fabrics, and other abstractions, such as axial planes of folds, should be marked on the map at the time of their recording. Attitudes of material planes (e.g. layering or schistosity), and abstractions (e.g. axial surfaces), are recorded as strike and dip. Directions of both linear elements (e.g. fold hinges) and the intersections of planar features (e.g. bedding/cleavage intersection) can be recorded as either bearing and

Profile in meters	Layer Thickness	Metamorphic Rock	Protolith (?)
360 m	6 m	Light quarzite with minor crossite	Clastic sediment (arenite)
350 m	10 m	Mafic schist (Qz, Ep, Amp, Chl.- Gar, Chl)	Volcanic
	4 m	Quarzite with dolomite pebbles	Reworked clastic sediment
340 m	4 m	Calcite marble with quartzite layers, strongly folded	Mixed carbonate and clastic sediments
	5 m	Calcite-Dolomite marble interlayers, weakly folded	Sediment with high carbonate content
330 m	4 m	Calcite marble, mouse gray, finely banded	Carbonate
	5 m	Calcite marble with Dolomite layers, not folded	Carbonate, Cc-Dol interlayering
320 m	7 m	Calcite marble, somewhat lighter than above	Carbonate
	2 m	Dolmite marble band (shows later shear fabric)	Carbonate, completely converted to dolomite
310 m			
300 m	30 m	Calcite marble, at top about 50 cm with dololmite intermixed then relatively homogeneous, finely banded, gray	Carbonate
290 m			
280 m	4 m	Mafic schist, weakly folded	Basic Volcanic
	1.5 m	Calcite marble lense, like the mable above, gray	Carbonate
270 m			
	25.5 m	Mafic schist, weakly folded (Crossite, Chl, Ep (Lin: 65/10), rare Garnet)	Basic volcanic, tuff
260 m			
250 m	2 m	Calcite marble, gray	Carbonate
	3 m	Mafic schist, similar to the overlying schist, 30 cm band of calcareous-mica schist/ conglomerate (like sample IP 48)	Basic volcanic with clastic carbonate detritus
240 m			
230 m			
220 m	60 m	Calcite marble, light gray, well layered (2-3 cm, 10-15 cm) at 15 m below the top calicte-dolomite marble interlayers of about 3 m show boudinage	Carbonate sedimentation with possible development of evaporite at the top
210 m			
200 m			
190 m	1 m	White mica schist layer	Clay-rich sediment, Distal carbonate
	1 m	isolated dolomite layer	Evaporitic carbonate
	2 m	Calc. mica schist/conglomerate layer, like above	Clastic sediment with high carbonate content
180 m	11.5 m	Mafic greenschist (rich in Ep, Gaucophane)	Basic Volcanic
170 m	2.5 m	Calcite marble band, gray/brown	Carbonate
160 m		Mafic schist, Probe analysis PR 2.1 (EP 251)	Basic volcanic
50 m		at 160 m above SL., transition to calc. mica schist (EP 81 and EP 82)	Transition from basic volcanic to clastic sediment with high carbonate content
40 m	170 m		
30 m			Clastic sediment fine grained to bis clayey with minor volcanic detritus
20 m			
10 m		Calcite marble	Carbonate
0 m			

Figure 2.8 *Detailed lithostratigraphic log of metamorphic units found south of Syringas, Isle of Syros, Greece. (Log courtesy of John Schumacher from thesis and mapping data compiled by Jörg Pohl, University of Freiburg 1999).*

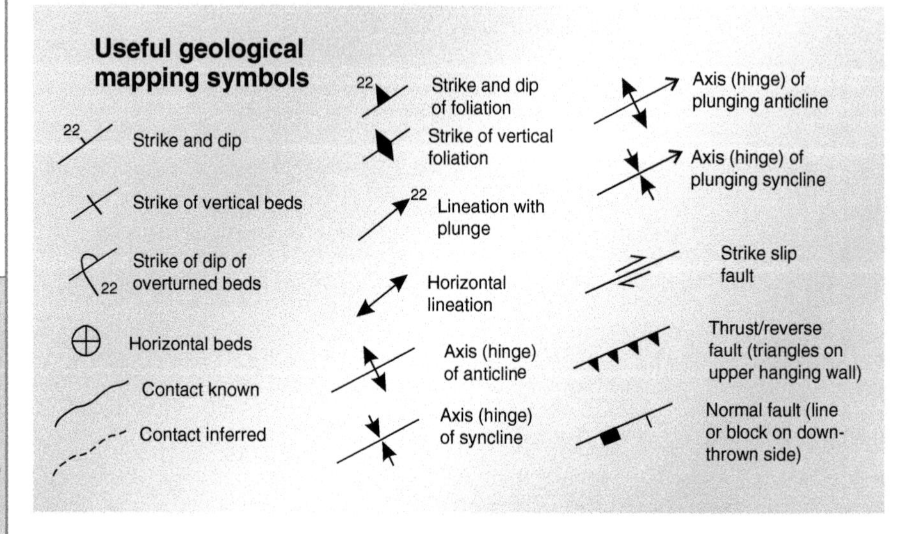

Figure 2.9 *Some of the common mapping symbols used in metamorphic terrains (see mapping guide in this series for further details).*

plunge, or as pitch on a measured plane. It is important to be aware of some of the key symbols that you require when recording the structures of metamorphic rocks in the field such as bedding, contacts, joints, and cleavage. A summary of some of the common symbols used in metamorphic-specific mapping is presented in Figure 2.9 (see also mapping and structural books in this series).

At complicated exposures, it may be required to map at a finer scale than the normal scale, which would be the basis of field slips. In this case, you might use detailed sketch maps in your notebook (e.g. Figure 2.10), create a high-resolution map from scratch using grids and cairns, or even mark out information on (or with overlays on top of) enlarged satellite maps of the region. In all cases you will be using the same style and symbols for consistency, and careful cross-referencing between your notebook and separate maps is essential. It is sometimes possible to attach the resultant map into the relevant part of the notebook; otherwise, keep it as a separate detailed mapping slip/example. Again, refer to the mapping book within this series for further ideas and pointers.

The field map is a working tool that shows the status of field evidence and helps in the synthesis of individual records into a three-dimensional picture of the rocks being studied. Contacts, faults, and markers are recorded as lines. The outcrop areas of different formations, and the exposures of these outcrops, are recorded by colour. Locations of all key localities are marked, as are the lines of logs or transects, of any areas studied in greater detail, of peculiar rock or key mineral occurrences, or of any samples collected. Field cross-sections are normally attached to the field map, to avoid discrepancies between the two, and to encourage on-the-spot three-dimensional thinking. The cross-section attached to the field map should show all contacts, representative traces of any banding and fabrics, traces of axial surfaces of upright folds and the senses and style of minor folds. Ultimately you will be aiming to work up your maps into a final metamorphic map of your area (e.g. Figure 2.11), and this final style will depend on the type of geology you are showing and the conventions you need to use for your particular project.

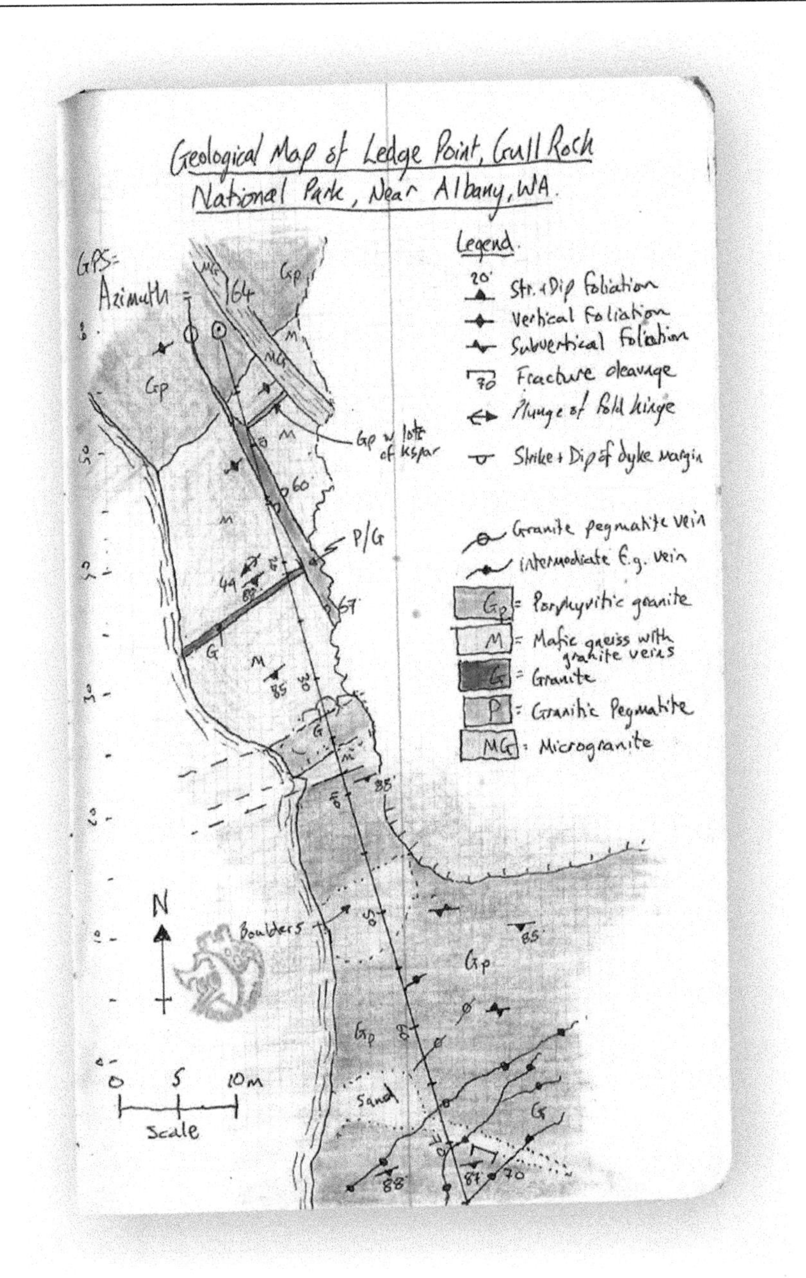

Figure 2.10 *Detailed map sketch within a notebook (courtesy of Nick Timms).*

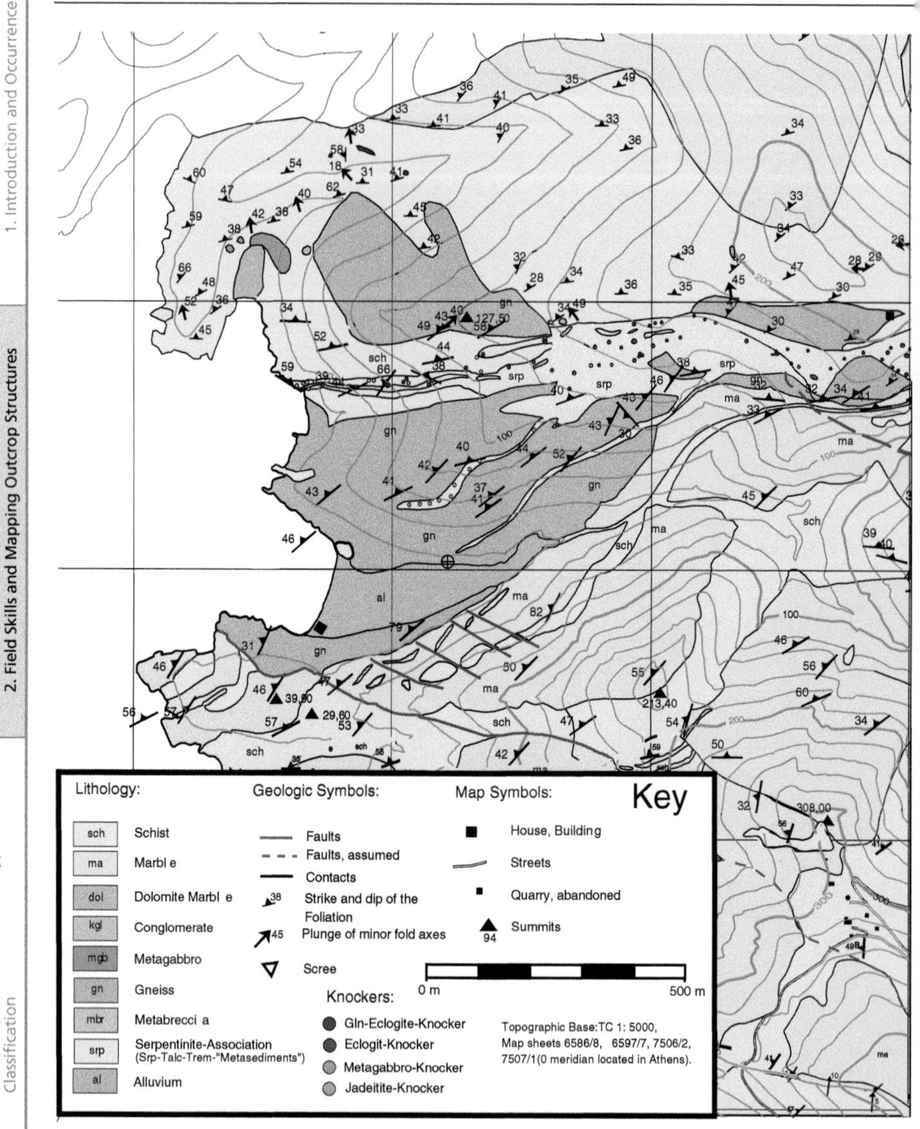

Figure 2.11 *Example of a metamorphic map of part of the Isle of Syros, Greece (Map data courtesy of John Schumacher from a compilation of University of Freiburg thesis mapping by Uwe Schmitt, Jörg Peters, Jörg Pohl, and Holger Schmolz, 1999–2000).*

2.3.4 Sample collection, naming, and recording

You may need to collect one or numerous samples during your fieldwork, with the objective of subsequent laboratory study. Assuming that you have the relevant permits for sample collection, you may be able to learn much more about the regional geology of the area by studying samples at the thin-section scale, or by undertaking subsequent geochemical or isotopic study. When collecting samples, it is always important to keep in mind this question: *what exactly are you collecting them for*? Studies aiming to decipher metamorphic pressure and temperature, deformation mechanisms, or geochronology all have different characteristics, require different sampled volumes, and may or may not require orientation information on the samples, so make sure that you are aware of what would be most useful for each branch of subsequent study *before commencing fieldwork.*

Sampling in the simplest sense is straightforward: use a hammer, a drill, or a saw to extract a large enough piece of rock for your purposes. This overlooks several key factors that you *must* consider before sampling, however: (i) Are you going to ruin an important outcrop by hacking it to pieces, and is this fair to others? (ii) Is there a less weathered part of the outcrop that would be more appropriate to sample? (iii) Have you found the part of the outcrop with the features that will be most useful for you (for example, the part with the most obviously interesting mineralogy)? (iv) Should you be just knocking off a piece of sample, or do you need to carefully annotate the specimen in the field, so that you have a record of its original orientation (which may be important if you want to reconstruct a deformation history)? (v) Can you safely extract a sample without exposing yourself to undue risk?

Once you have an appropriately sized sample, it is important to make a record of it in your notebook. Each sample should have a unique number, and you should never use that number again. A good way of doing this is to combine a name for the field campaign with the year of that work and a unique identifier for each sample collected in that field campaign, so that the twenty-third sample collected in a field campaign on the island of Syros in the year 2018 might be SYR/18/23. If your field campaign involves multiple scientists accumulating a single sample collection, you might also incorporate who collected each sample, so that Jerram or Caddick's samples 23 and 24 might be labelled SYR_J/18/23 and SYR_C/18/24, respectively. You can choose to format this sample label however you want, but be consistent and make sure that all sample names are unambiguous. Your sampling numbering strategy should be described somewhere in your notebook for future reference.

Once you have a sample number, write it directly onto the sample with a marker pen, unless that ink could compromise a geochemical study that you plan. Write the number again on a scrap of paper and then put the paper and sample into a sample bag, writing the sample number on that bag too. If you need to ship samples home, you may eventually need to wrap them carefully to avoid breakage. Every sample that you take should be associated with an entry in your notebook, so that samples can be put back in the context of outcrop-scale observations and hypotheses at a later date. It is often useful to briefly describe in your notebook the rationale for taking each sample, for example describing that a sample was taken for more detailed thin-section-scale characterisation of mineralogy, because you think it might be particularly useful for detailed geochemistry or geochronology, or because you suspect that it might reveal important structural information at the sub-cm-scale.

2.3.5 The nature of particular contacts

Mapping demands that you trace contacts, follow marker horizons, and so on. You will need to define the main rock formations and any other key units within these (e.g. Members), and more information about this is provided within the reference sections in Chapter 8. Description requires you to record the formation that lies between such contacts and markers, and this information may be more appropriate in either the notebook or sometimes directly on the map. These tasks are complementary. The only way to establish the uniqueness and significance of markers is by studying the formations in which they lie. On the other hand, every contact represents a relationship of some kind between the rock formations or units it separates. Ask questions about the geology of the area, and you should find that to answer them you continually shift between mapping and describing. Don't forget that in metamorphic terranes you are likely to find primary (lithological) boundaries that are overprinted by varying intensities of metamorphism, and both of these pieces of information might be crucial to your success.

Contacts between metamorphic rocks can be ambiguous and potentially difficult to deal with. They may be pre-metamorphic, as original contacts in the protolith (e.g. depositional surfaces in sediments, intrusive contacts, or originally faulted). In other cases they represent a change in metamorphic grade, either discrete or gradational, produced during metamorphism and associated deformation. In this case, the style of the metamorphism can vary between contacts, representing different grades and processes that have affected the rocks (e.g. metasomatism, abundance of minor intrusions, finite strain, intensity of deformation fabric, and structural style). It is important to be aware of compositional differences in the rocks, as some compositions will determine the type and abundance of metamorphic minerals present (see Chapter 3 and reference Chapter 8). You should also note where deformation has been unevenly distributed or partitioned, or where rocks of very different metamorphic grade or known ages are next to each other in the field (e.g. Figure 2.12). Also, the boundaries of a metamorphic rock mass with non-metamorphic rocks should only be interpreted after a full appreciation has been gained of the nature of the metamorphic rock mass's internal contacts. To this extent you may be overlaying metamorphic grade contacts, for example, onto a map outlining the main formations in the area, and these may or may not be coincident (think of marking Barrovian zones or metamorphic aureole zones on a map in cases where they overprint different rock formations).

Figure 2.12 *Example of contact relationships within a complex terrain, including unconformable relations, intrusive contacts, and contact metamorphic aureoles. Within the metamorphic suite at the base, complex internal contacts also exist. You will possibly need to decide how to best deal with and to characterise all of these types of complexity in both your notebook and on a map (photo Dougal Jerram).*

It should be standard practice to walk along contacts where possible, looking to each side to try to detect:

1. Angular discordances, which turn into the line of the contact.
2. Changes in strain, metasomatic alteration, or retrograde metamorphism towards the contact.
3. Faults belonging to the same system and/or in the same orientation as the local contact, to which the main displacement or contact may transfer further along its length, leading to a change of the apparent thickness of units.
4. Any features that have been offset, either by a straightforward fault or by a broader zone of ductile displacement.

Remember that rocks are three-dimensional, and banding, lithological, or structural contacts consist of planar surfaces, even if they appear as lines on two-dimensional outcrop exposures. These surfaces may not be perfect, simple planes, and their undulation might reveal important information about the nature of successions that have been metamorphosed or the intensity and direction of transport due to faulting. Wherever possible, therefore, banding should be examined on surfaces that reveal a variety of orientations, and any variations in properties within the third dimension should be reported.

2.4 Digital 3D Outcrop Mapping

In recent years, it has become easier to use technology within geological fieldwork. This can take the form of using tablets with mapping apps/software, or the use of drones to create 3D virtual outcrop models. Much of this chapter has been aimed at advising you in terms of creating and recording excellent notes, which separate observations from interpretations within a conventional notebook, and the mapping of units on conventional maps. In the case of apps and software that can be used within tablets, *most of the basics are the same*: high-quality observations and methodical recording of observations, measurements, and interpretations, though with digital forms or templates to help with consistent recording of data. It might be useful to initially learn to record observations on paper, for instance in an undergraduate mapping dissertation, transitioning towards digital mapping techniques at a later date (e.g. in graduate studies or as a professional geologist). Times are changing rapidly though, and it may be the case that you are able to incorporate digital advances directly into undergraduate studies. Again, the crucial element is for you to develop a systematic workflow that allows you to work carefully and methodically, and to record observations, interpretations, and samples clearly and consistently.

The advent of cheap, high specification drones has, in many ways, revolutionised the way that we can access and map areas. They can be of particular use to gain insights into poorly accessible areas, and to visualise large structures and contacts. 3D virtual outcrops are derived from drones using overlapping orientated photographs, which can be used to construct a surface representation of the outcrop, coloured according to the photos taken (Figure 2.13). Computer packages such as Agisoft's Metashape enable the geometrical calculations and surface reconstructions, and additional software (e.g. the Virtual Outcrop Geology Group's LIME) can be used to visualise and measure information from the virtual outcrop. Figure 2.14 shows an example of highly deformed folded schists in the Ugab Region of Nambia, as one such example of a virtual geological outcrop. The use of 3D mapping is increasing, and it can be considered a useful tool that can accompany a detailed field campaign and can provide additional help back in field camp or the lab. The virtual outcrop models can be further set up to include your data as virtual fieldtrips and are opening the way for better inclusivity for people with limited access to physical outcrops. A database of accessible 3D outcrops has been compiled (http://v3geo.com), with examples being added regularly.

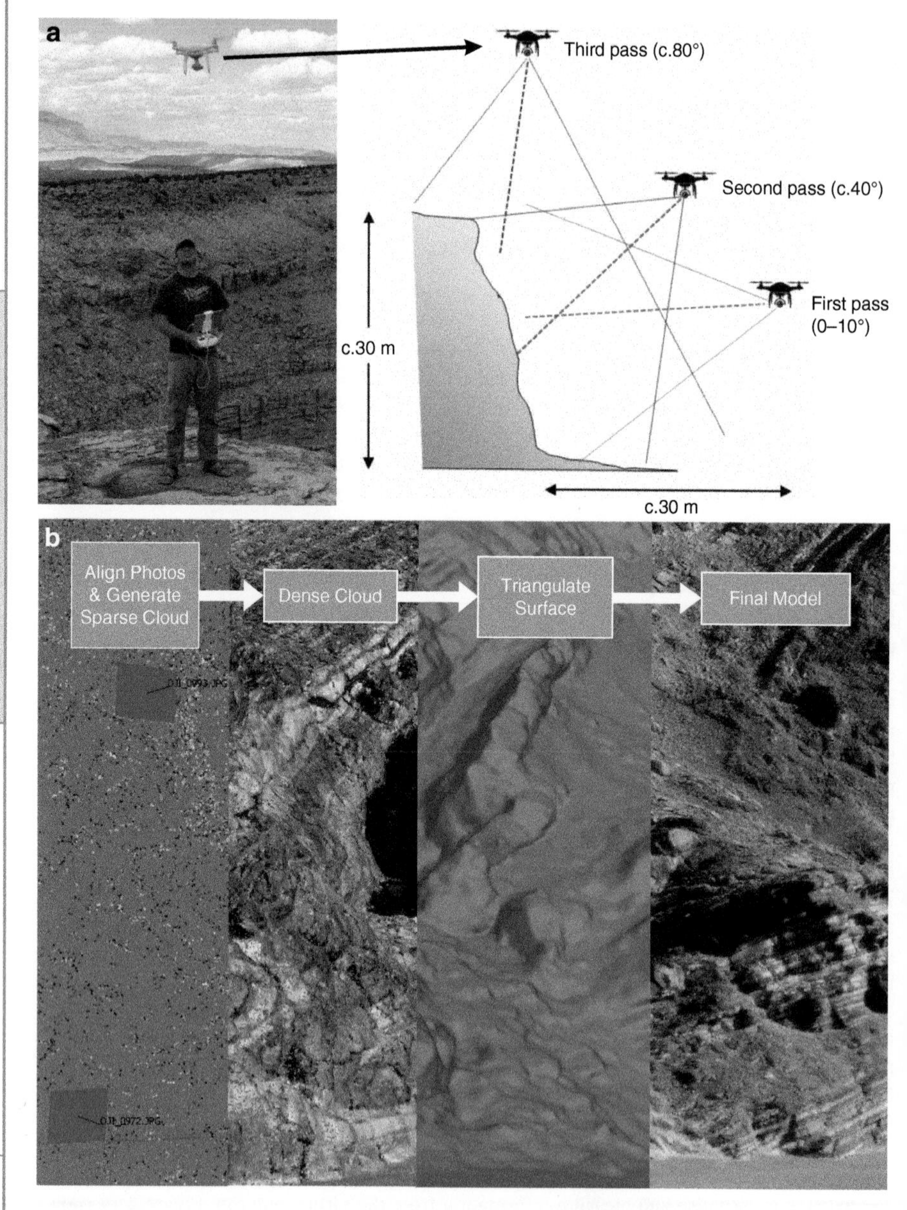

Figure 2.13 *(a) The process of mapping outcrops in 3D using a drone. The correct amounts of overlapping photos are required from different positions. (b) These can then be input into software to calculate a 3D point cloud, construct a 3D triangulated surface, and finally produce a final 3D outcrop model (courtesy of John Howell).*

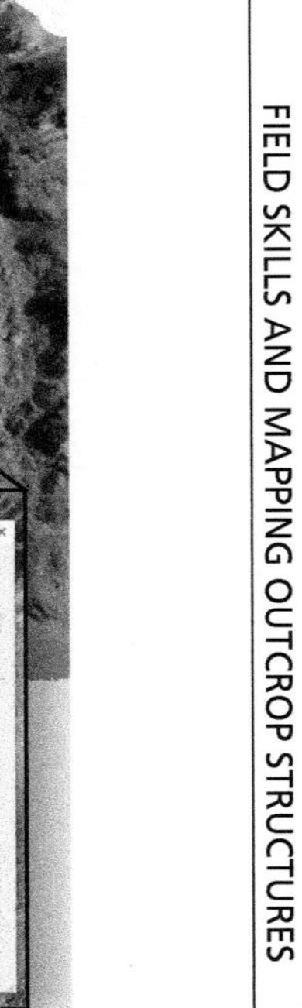

Figure 2.14 *3D outcrop model of large folds in metamorphic calc-schists, NW Namibia. Insert shows measurement of fold axial plane on inaccessible cliff made within LIME software (3D model courtesy of John Howell).*

METAMORPHIC MINERALS, ROCK TYPES, AND CLASSIFICATION

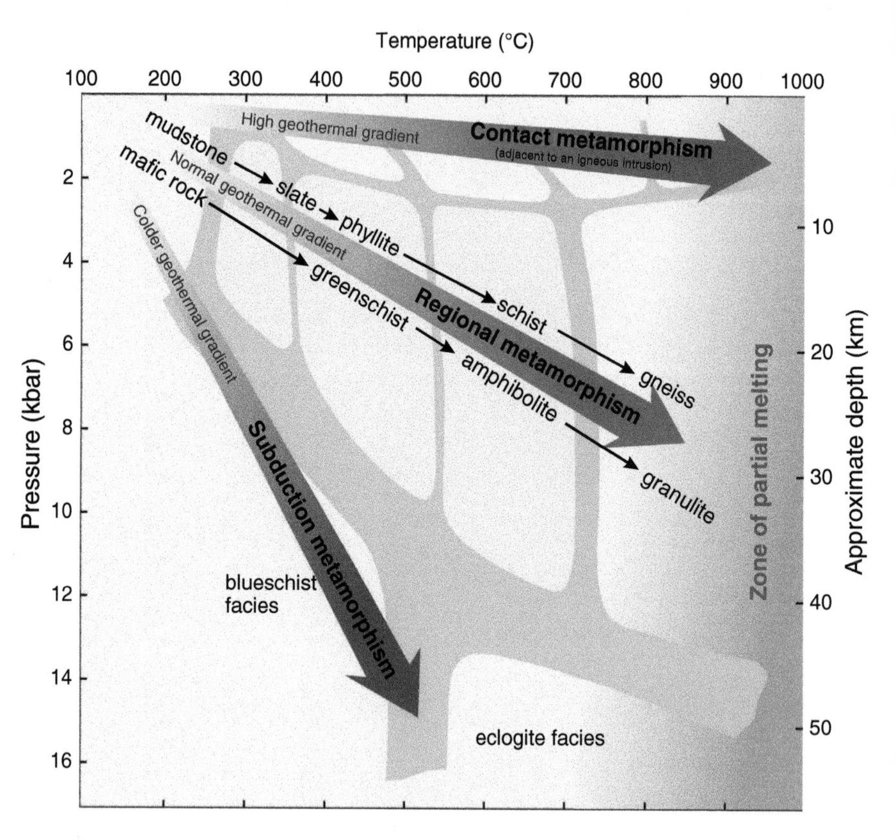

The building blocks of metamorphic classification are found in the relationships of pressure and temperature.

3

METAMORPHIC MINERALS, ROCK TYPES, AND CLASSIFICATION

In this chapter, we will consider the basic requirements for key mineral and rock type identification that will help you name and classify your metamorphic rocks. The matters considered in this chapter provide the basis for every description of metamorphic rocks, and the chapter should be used in close conjunction with the additional reference tables and diagrams in Chapter 8. The first stage is to consider the visible mineral types in the rock (Section 3.1), because the effective identification of key metamorphic minerals and assemblages can help determine metamorphic grade and the appropriate nomenclature of your rock. Next, we look at the main metamorphic grades in terms of pressure and temperature, discussing how rock types and some key minerals will be present or will change at different conditions (Section 3.2). This section is colour coded to help you relate between the tables and diagrams. We can then consider the naming of rock types (Section 3.3), and the general plan that should be followed when describing them (Section 3.4). The final section (Section 3.5) considers deductions about compositional category and metamorphic grade that may be justified based upon field observations. Tables of detailed information relevant to these sections are given at the end of the book (Chapter 8) for easy reference in the field, with some pertinent details reproduced in this chapter also.

3.1 Minerals

Wherever minerals are visible and identifiable, they provide the basis for rock-type names, used for immediate field notes and for full rock-type descriptions. It is from these descriptions that most statements of origin and metamorphic grade can be made. Only if minerals are not identifiable will a more generic description use physical properties, such as colour, as the basis for rock-type recognition. We focus initially on the types of mineral and mineral associations, because certain key minerals are found at different metamorphic conditions (see Figure 3.1). To help with the identification of some of the key metamorphic minerals, Chapter 8 contains an extended table of common mineral types with some of their main properties. In both cases, we reference these minerals to the tables presented in Section 3.2, which will help you start to build up a knowledge of what types of metamorphic rocks they are most generally found in and what they reveal about metamorphic grade and rock composition.

3.1.1 Recording assemblages, proportions, and properties

Field notes should be made in such a way that deductions about rock composition (protolith), evolution, and metamorphic grade can be produced as easily as possible, ideally immediately at the outcrop but potentially after further work on additional outcrops. Deductions about *grade of metamorphism* may be based on the following:

The Field Description of Metamorphic Rocks, Second Edition. Dougal Jerram and Mark Caddick.

1. The *existence* of a particular diagnostic mineral, as in the classic 'Barrovian sequence' for pelitic rocks (see Figure 3.1).
2. The *coexistence* of several minerals, making up either part of or the entire group of minerals (*the assemblage*) that constitutes the rock.
3. The *absence of minerals* or the *absence of combinations* of minerals that might have otherwise been expected.

It should be noted just how important it is to know *which minerals occur in contact with each other and which minerals do not* (even if those minerals do occur in the same outcrop or hand specimen, they may not be equilibrated with each other if they are never found in contact). It is always important to keep in the back of your mind the concept of what the original protolith rock composition might have been, as in many cases the richness of the chemical variation, or lack of, in a rock will determine whether metamorphic minerals will grow at all. To this extent we have categorised the different types of protolith using the letters A–G to describe the following types (see also Table 3.1): A – Hydrated Ultramafic; B – Mafic Igneous; C – Felsic Igneous; D – Psammitic Sediments; E – Pelitic Sediments; F – Carbonates; and G – Semi-Carbonates. The general character of your protolith will fit somewhere into these broad categories, and you may also be able to further ascertain the types of protolith by finding structures that reflect original characteristics of the non-metamorphosed rock type (for example, cross bedding and grading in sediments, pillow structures

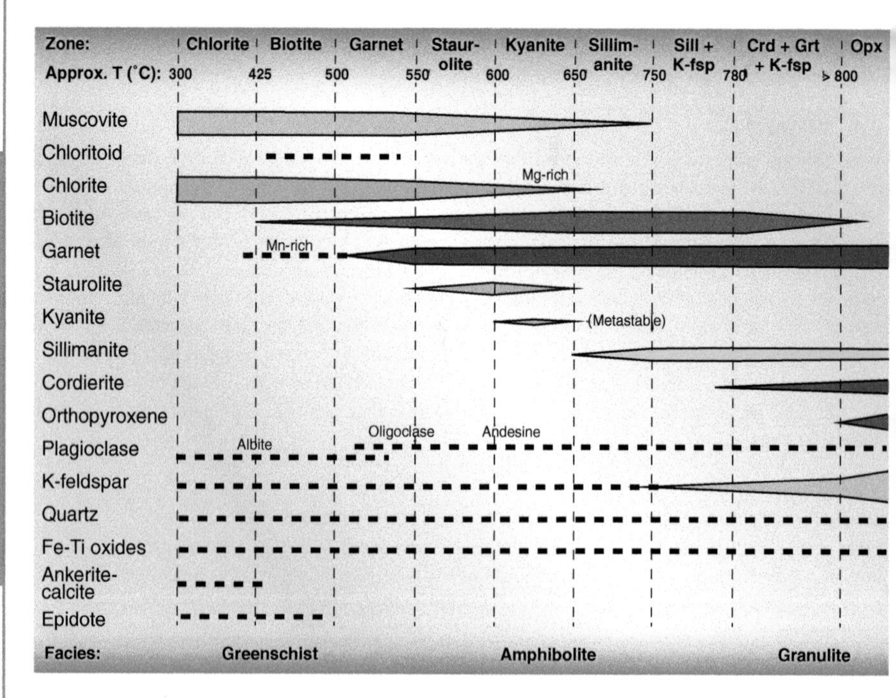

Figure 3.1 *Generalised mineral occurrences and compatibilities in the typical Barrovian metamorphic zones in pelites in a normal (regional) geothermal gradient. Some of the key minerals are coloured with a schematic indication of abundance (thickness of band scales with abundance of mineral). Dashed lines show possible mineral occurrence with no indication of relative modal abundance (adapted from Best 2003).*

Table 3.1 An overview of the key rock types formed from the major protolith groups upon progressive metamorphism.

	Grade code	A) Hydrated Ultramafic	B) Mafic igneous	C) Felsic igneous	D) Psammitic Sediments	E) Pelitic Sediments	F) Carbonates	G) Semi-carbonate	Grade code
		General Classification of Protolith Type							
Unmetamorphosed		Serpentinite	Basalt	Rhyolite/ Granite	Sandstone	Shale	Limestone/ dolomite	Marl or mixed protolith	
Increasing metamorphic temperature at crustal and thickened crustal depths →	VLG		Zeolite	Little change without fluid addition or deformation →	Increasing grainsize Increasing bulk hardness →	Slate	Increasing grainsize Increasing bulk hardness →	Calcareous Slate	VLG
	LG	Talc schist	Prehnite/ Pumpellyite			Phyllite		Increasing grainsize, isotropy & bulk →	LG
	MG1	Amphibole Peridotite	Greenschist			Schist			MG1
	HG1		Amphibolite			Gneiss			HG1
	VHG	Peridotite	Granulite		Quartzite	Migmatite	Marble	Diopside-gneisses	VHG
Increasing temperature at high pressure →		Serpentinite	Basalt	Rhyolite Granite	Sandstone	Shale	Limestone/ dolomite	Calcareous Slate	
	MG2	Chlorite-Peridotite	Blueschist			Micaceous Blueschist		Calcareous Blueschist	MG2
	HG2	Peridotite	Eclogite		Quartzite	Micaceous Eclogite Gneiss	Marble	Calcareous Eclogite	HG2

Colour categories for each protolith are used elsewhere in this book, as are the grade codes. These codes are very low grade (VLG), low grade (LG), medium grade at low to intermediate pressure (MG1), high grade at low to intermediate pressure (HG1), very high grade (VHG), medium grade at high pressure (MG2) and high grade at high pressure (HG2).

in mafic igneous, etc.). The preservation of such 'original' structures will vary with the extent of grade and deformation that the rock has undergone (e.g. Figure 3.2), but an understanding of the types of structures that you might have expected the original rock to have can be very helpful and will be touched on in more detail in Chapter 4.

When describing your protolith you may even consider the term 'fertile', in which some protolith compositions will naturally tend to form greater proportions of certain minerals at a given pressure and temperature, or may grow those minerals at lower temperature than less fertile compositions. For example, pelitic rocks (meta-shales) tend to contain more staurolite if they are relatively

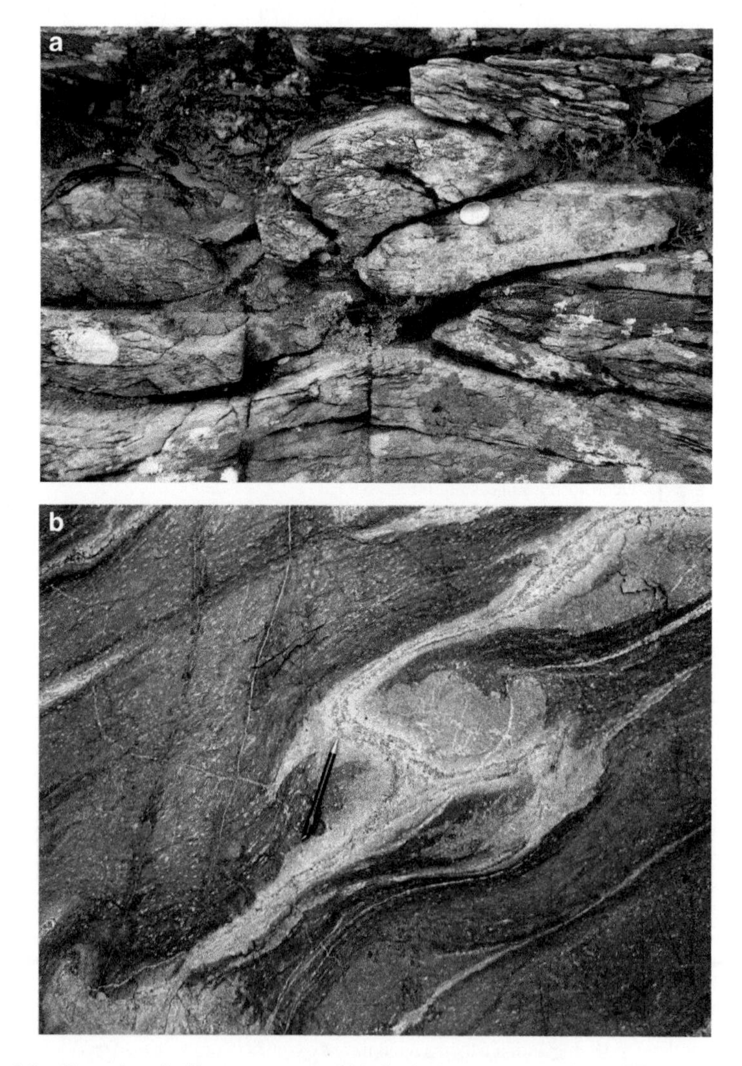

Figure 3.2 *Examples of pillow structures within basic igneous protoliths at different metamorphic grades. (a) Clear pillows in almost pristine greenschist facies pillow basalts, Western Norway. (b) Deformed pillow basalts with epidotised pillow margins, Matchless Amphibolite of the Damara Orogen, Kuiseb Canyon, Namibia (photo (a) Dougal Jerram, photo (b) courtesy of Susanne Schmid).*

aluminous, will grow garnet at lower temperatures if they contain more manganese, will begin to melt at a lower temperature if they contain or are fluxed with a hydrous fluid, and will generate proportionally more melt if they contain more of this H_2O. You cannot easily determine the exact aluminium, manganese, or H_2O contents of a metamorphic rock in the field, but some understanding of these trends will help to explain, for example, particularly staurolite-rich rocks, or rocks in which staurolite appears to have grown before biotite.

However, for determining metamorphic grade, the absolute proportions of coexisting minerals are far less important than simply identifying what minerals are present and, to the extent possible, which minerals are found in contact with each other. That being said, a fairly close approximation to the rock's *bulk chemical composition* may be made by summing the compositions of constituent minerals, weighted in their correct proportions. The exact compositions of the constituent minerals cannot generally be determined in the field, and most minerals will contain trace impurities or other complexities, but having a good knowledge of the idealized compositions of minerals is very helpful. For example, knowing that diopside will have a composition very close to $CaMgSi_2O_6$, tremolite will have a composition close to $Ca_2Mg_5Si_8O_{22}(OH)_2$, garnet will have a composition close to $X_3Al_2Si_3O_{12}$ (where X can any combination of Fe, Mg, Ca, and Mn), calcite and dolomite will have compositions close to $CaCO_3$ and $CaMg(CO_3)_2$, respectively, and quartz will have a composition close to SiO_2 can be very helpful in determining the appropriate protolith of any rock with several of these minerals. In many cases, bulk-rock composition is the main clue to the nature of the rock before metamorphism, and this can be geologically as, or more, important than the metamorphic conditions. Therefore, *mineral proportions should be recorded where possible.*

Clearly, the recording of minerals, their mutual and mutually exclusive relationships, and their proportions is a time-consuming and potentially tedious task. However, in addition to justifying deductions about metamorphic grades and pre-metamorphic rock types (e.g. Sections 3.3 and 3.4), this information may contribute to a synthesis of the rock mass description, including description of the relationships between formations (see Chapter 8).

In some cases, you will need to report *properties of particular minerals*, as well as relationships between minerals. This is true if any mineral's properties are unusual, if they cast doubt on the validity of that mineral's identification, or if the mineral is one which often varies in its properties. For example, identification of quartz with a grey-blue colour should be recorded, because it might hint at (though not conclusively) a high temperature of metamorphism or sediments that were derived from high-temperature rocks. Likewise, if some garnet crystals in a sample are bright orange while others are dark red, this should be recorded (even if you cannot at that moment interpret why this is the case). To say that quartz looked glassy, or that calcite effervesced in hydrochloric acid, or that a mica could split into cleavage flakes is, however, a waste of time – those are fundamental properties of each of those minerals and can generally be taken for granted. As a rule:

1. If a mineral is *recognized*, record it and its proportions.
2. If a mineral is *suspected*, or an identification *doubted*, report why.
3. If a mineral is known to belong to a certain *group*, or to have a *compositional range*, record those characters which are variable for the group or range.
4. If you cannot decide what a mineral is, describe it fully.

3.1.2 Mineral identification

Mineral grains that are visible with a hand lens should be identified, or otherwise narrowed down to a particular group with similar properties. The list of minerals given in Chapter 8 and summarised in Figure 3.1 is based on their normal characteristics, of which colour and shape ('habit') are not always reliable. There are many potential variations which you must beware of, but this complexity can turn to your advantage when interpreting protolith type and metamorphic history. For example, metamorphic amphiboles may occur with several potential colours, which may be indicative of metamorphic grade or protolith compositions:

1. Hornblende is likely to be black or to have a slight green shade, and is likely indicative of 'amphibolite-grade' conditions experienced by mafic igneous and some sedimentary protoliths.
2. Glaucophane will have a distinctive blue colour and forms during high-pressure metamorphism of mafic igneous and some sedimentary protoliths.
3. Tremolite will be a creamy white, tending towards a dark green as iron is substituted for magnesium. It commonly results from contact metamorphism of siliceous sedimentary rocks with abundant calcium and magnesium and can also be found in greenschist facies metamorphic rocks derived from ultramafic or carbonate protoliths.

Many variations on these colours are possible, due in large part to the fact that amphiboles can have a wide range of possible compositions. In addition, the habit of amphibole can also vary significantly, occurring as:

1. Short stubby grains in direct replacement of pyroxene crystals.
2. Needle-like or brush-like sprays if grown in a schist after deformation.
3. As equigranular grains in an amphibolite which has accommodated continuing ductile strain.

Many other habits are possible, but description of amphibole as being in just one of these three groups is useful. Likewise, sillimanite can occur in tabular or fibrous forms (e.g. Figure 3.3), in this case potentially revealing something about metamorphic grade or growth reaction.

Descriptions of *exsolution, twinning, alteration*, and *weathering* can all be useful, as can colour. For example, the bronze colour of bronzite is distinctive, as are the golden-brown and caramel colours of partially weathered biotite. Igneous pyroxenes often possess one main parting along exsolution lamellae, which can give them a micalike appearance, except that they will break (with difficulty) into small angular granules instead of splitting into flexible flakes. The surface of a large, weathered carbonate crystal can look furry, due to the etching of traces of sets of intersecting cleavages or twin lamellae. An orange or rusty deposit on or around weathered carbonate grains indicates a ferroan variety such as siderite (e.g. Figure 3.4). A surface tarnish or stain on exposed rock surfaces is often indicative of the weathering of sulphide minerals such as pyrite, with the stain deeper on more sulphidic regions and around individual sulphide grains. Other examples of colour variations may result from local weathering or hydrothermal conditions, and it can be worthwhile to draw up a list of such features for a field area.

It is rare for every grain of each mineral in a rock to be individually identifiable. Most minerals are instead identified by constructing a list of the features visible in a number of grains, assuming that these grains are already known to be the same mineral. This highlights the extent to which human vision may perceive likenesses and differences despite an inability to identify minerals. This ability should be used consciously to *extrapolate* identifications from one rock to another. For example:

1. A mineral may be *idiomorphic* (having its own shape), and identifiable when surrounded by one set of minerals, and clearly identical in colour and weathering characteristics to shapeless grains elsewhere.
2. Grains may be larger in some patches, such as veins, than elsewhere, and so display features like cleavage more clearly. But other characteristics might suggest that the smaller grains are of the same mineral.
3. Idiomorphic vein minerals may be crystallographically continuous overgrowths on smaller shapeless grains at the margin of the host rock.

Whenever mineral identification relies on extrapolation, this should be noted at the time and reported. Without this, false correlations may later not be identified as such, leading to misleading conclusions.

Figure 3.3 *Sillimanite in two forms: (a) Folded fibrolitic sillimanite from the Beartooth Mountains, Wyoming, and (b) coarse tabular sillimanite crystals from the Pikwitonei Granulite Domain, Manitoba (photos from Victor Guevara).*

Accessory minerals (those in quantities so small as to be neglected in considerations of rock names) should also be reported wherever possible. It is good practice to search for holes on weathered surfaces, where soluble minerals may have been leached out, and to search in veins and coarse-grained patches. If accessory minerals are found in these regions, they are likely to also occur in the rock as a whole. If these minerals might be of later importance, for example if they will eventually be used for isotopic dating of the metamorphism, samples will probably need to be taken for subsequent, detailed thin-section scale petrography.

As discussed earlier in the example of amphiboles, *colour* can both be useful for identifying minerals, and be problematic and potentially misleading. For example, many of the minerals in metamorphic

Figure 3.4 *Vein containing white calcite ($CaCO_3$) crystals that have orange-brown staining associated with weathering of siderite ($FeCO_3$) crystals (photo Mark Caddick).*

rocks can be found in various shades of green. However, with some practice, the specific greens of jadeite, actinolite, olivine, serpentine, chlorite, epidote, and prehnite can all be distinguished in the field. If possible, we suggest learning how to recognize these and any other minerals that you anticipate finding in a field area before embarking on your fieldwork, and potentially carrying representative samples until you are confident with identification. Though some colours are true mineral colours, others are due to the presence of small inclusions (e.g. whitish sericite or reddish iron oxide or blackish ilmenite in feldspars). Other minerals can have misleading colours due to the presence of other trace impurities, either in solution within the mineral or as inclusions that are far too small to identify in the field or even with an optical microscope (e.g. typically colourless quartz, SiO_2, can turn a characteristic blue-grey colour if it incorporates sufficient TiO_2). At other times, the apparent colour of a transparent grain may actually be that of the surrounding material (e.g. green imparted to quartz-feldspar rocks by interstitial phengite or sericite). It is important to note that the colours quoted in many books are for ideal specimens or for minerals in a thin section. A mineral with bold colour in a thin section will most likely be black to the naked eye (e.g. hornblende, biotite, or spinel), whereas a mineral with colour to the naked eye may be pale to colourless in a thin section (e.g. garnet, staurolite, and kyanite). True mineral colours are imparted by certain elements only. Minerals containing only elements with atomic numbers up to that of calcium, or these elements plus alkali metals or alkali earths, are generally colourless. Minerals having transition elements or heavy metals as essential components (roughly speaking, mafic and heavy minerals) are dark. A small amount of transition element substitution into pale minerals gives bright colours (e.g. small amounts of iron in olivines, phengitic micas, tremolitic actinolites, epidotes, and prehnites). A duller effect is usually produced by finely disseminated pigments (e.g. ferric oxides) in a pale mineral base. A few elements are associated with a particular colour (e.g. chrome with green, which gives pyroxene the bold green colour seen in Figure 3.5a). In this context, the ochre colours of ferric oxides and hydroxides are abnormal. Ferric iron is often associated with green in silicates (e.g. epidote, Figure 3.5b) so green is not always an indication of chemically reduced condition.

Basic mineral *symmetry* may be much more clearly apparent from cleavages than from *shapes*. However, grain shapes may be considered in conjunction with other features. For example, a circular-sectioned, rod-shaped mineral probably has more than one symmetrically equivalent direction

Figure 3.5 *(a) High-pressure metamorphic pyroxene, with green colour due to the incorporation of chromium. (b) Epidote in a glaucophane–garnet schist, where the green colour is due to iron. Both photographs from Syros, Greece (photos Mark Caddick).*

in the plane of the circular section, and so belongs to an optically uniaxial system. However, if it possesses just one cleavage along its length, such symmetry is refuted, and the mineral must have orthorhombic or lower symmetry. Such quick checks can prevent a number of misidentifications. Where evident, twinning can be useful, with twinning in minerals such as staurolite relatively uncommon but helpfully diagnostic in many cases.

Cleavage, and particularly the angle between cleavages, may be observable on a roughly broken mineral surface that cuts the cleavage orientations. The surface consists of numerous minute steps along the cleavages, which, though they may be invisible, can all still reflect the light in the same directions. Given a directional light source, such as the sun, it may be possible to 'catch the light' off different cleavage planes and to make a rough measurement of the angle between them. In particular, this should be a standard approach for distinguishing between otherwise similar amphiboles and pyroxenes.

Despite all this, the greatest problem with minerals is often seeing them at all. Some minerals of below or above average resistance to solution weathering may show up on weathered surfaces. Others may be distinguishable by colour only when fresh. Some may show on lightly weathered surfaces by etching of cracks around their grain shapes. The roughness of dry rock may scatter light and show up some grains. Others may be more visible on a wet surface, which scatters light less. All of these things depend on local conditions. The only rule is to keep trying, with a hammer, hand lens, water, hydrochloric acid, and a streak plate. Of these, your hand lens and some patience may be your most important tools.

3.2 The Basic Classification of Metamorphic Rocks in *P-T*

Changes in the character of a rock as it is traced towards a heat source in a metamorphic aureole or through a larger domain that has experienced regional metamorphism are called changes in *grade*. Many changes occur upon increasing grade, including changes in grain size and the spatial organization of crystals, but changes in the assemblages of minerals that coexist in a rock are the most diagnostic of the pressure and temperature conditions that were attained. In the simplest form we can consider the main changes in the grade of a rock by considering the pressure and temperature *P-T* over the different rock types (e.g. Figure 3.6 and Table 3.1).

Most famously, this gave rise to the Barrovian sequence, in which the initial presence of the indicator minerals chlorite, biotite, garnet, staurolite, kyanite, and sillimanite were mapped and used as proxies for increasing grade in pelitic rocks (e.g. the zones in Figure 3.1 also shown in Figure 1.3). In most cases, increases in grade correspond to increases in temperature of metamorphism, but this does not mean that *grade* and *temperature* are synonymous. The same transition from one set of minerals to another may occur at different temperatures in rocks of only subtly different composition. However, we cannot 'see' metamorphic temperatures, so the transition of minerals is the best information we have for saying that different rocks have something in common, i.e. they have the 'same grade'. In other words, *pressure and temperature are measures of conditions that cause metamorphism. Grade is a measure of the effect* (metamorphism). Many techniques allow the metamorphic petrologist to quantify metamorphic *P-T* conditions, but these invariably require laboratory analysis rather than simple field observation. Understanding of what groups of minerals are generally associated with specific conditions, and an ability to recognize those minerals in the field, remains invaluable.

Mineral reactions differ in their response to *pressure*. Generally, at higher pressure, a mineral reaction will also occur at higher temperature, but the extent of this is different for each reaction. For example, the breakdown of feldspars and growth of clinopyroxene is a shallow line in *P-T* space (see the transition between 'granulite' and 'eclogite' in Figure 3.6a), but the transition from chlorite-dominated to amphibole dominated is a steep line (the transition between 'greenschist' and 'amphibolite' in Figure 3.6a). In either case, the transition from one grade to another cannot uniquely define the pressure of metamorphism unless the temperature is also defined.

Before we further consider the way in which we name metamorphic rocks, this is a good place to start to build a basic understanding of the common classification names and how they are associated with metamorphic grade, with reference to some pressure–temperature (*P-T*) diagrams. Here, we aim to initially be quite simplistic, leading you to a deeper knowledge as you progress. The 'facies scheme' that we have already seen in Chapter 1 (see Figures 1.2), reproduced in modified form in Figure 3.6a, is a useful starting point to consider the way in which mafic rocks will change with pressure and temperature to produce some of the classic metamorphic rock types. If we expand this concept across a variety of protolith types (defined as the original rock composition and type that has subsequently been subjected to metamorphism), and we simplify metamorphic grade to a series of general domains (e.g. very low grade, low grade, medium grade), the generalized rock types that

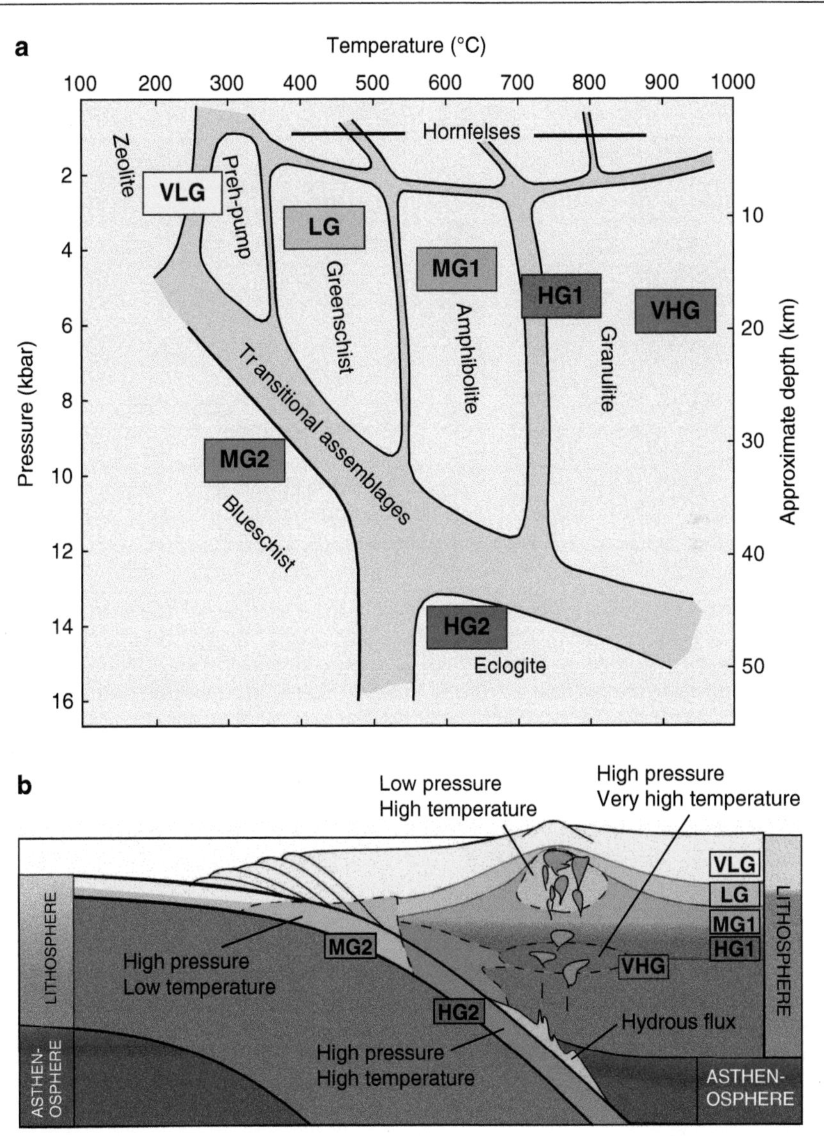

Figure 3.6 *(a) A version of the classic metamorphic facies diagram, showing the major rock types formed from mafic rocks as a function of pressure and temperature, and the general grade indicators used in this book. (b) Likely tectonic environments in which these conditions may be achieved (see also Figure 1.9).*

might be found can be listed and compared, as shown in Table 3.1. It is useful to be able to separate consideration of metamorphic *P-T* conditions from the rock protolith, and to recall that specific *P-T* conditions may be directly related back to tectonic setting or other metamorphic environment (e.g. Figure 3.6, inset figure at the start of this chapter, and Figure 1.2).

In the absence of substantial fluid input or output, the protolith plays the dominant role in controlling the chemical nutrients that are available for the growth of metamorphic minerals, and thus what

1. Introduction and Occurrence
2. Field Skills and Mapping Outcrop Structures
3. Metamorphic Minerals, Rock Types, and Classification
4. Understanding Textures and Fabrics 1: Banding, Cleavage, Schistosity, and Lineations

Table 3.2 *Some metamorphic rocks and the most diagnostic minerals associated with them.*

Rock-type Name	Diagnostic Minerals	Category (see Table 3.1 & chapter 8)
Amphibolite	Hornblende + plagioclase	B
Anorthosite	Plagioclase (calcic)	B
Blueschist	Any assemblage including blue amphibole	H (for H see chapter 8)
Eclogite	Omphacite + garnet	H (for H see chapter 8)
Granulite	Any assemblage including ortho-pyroxene, though clinopyroxene can also occur	CE AB
Greenschist	Albite + epidote + either actinolite, or chlorite, or both	B
Quartzite	Quartz as the dominant phase	D
	Carbonate minerals	FG
Marble Micaschist	Micas (often with quartz or carbonate or both)	CDEG
Peridotite	Olivine	A
Serpentine or Serpentinite	Serpentine minerals	A

See also Chapter 8 for an extended version with likely accompanying minerals in each rock type.

a metamorphic rock will look like, how it will deform, and what it should be named. With this in mind, it is important to consider some of the main rock types, and how they change when subject to different metamorphic conditions. Tables 3.1 and 3.2 consider simplified starting protolith categories, for instance grouping compositions into 'mafic igneous', 'pelitic', or 'carbonate', outlining some of the changes in rock name (Table 3.1) upon heating and some of the diagnostic minerals within these rocks (Table 3.2). In each case, labels A–G are used to signify the major compositional category, with label H used for primarily mafic rocks at high pressure (e.g. subduction zone) conditions. The approximate grade of metamorphism is denoted by the abbreviations (see also Section 8.1):

VLG - very low grade, LG – low grade, MG1 – medium grade (regional), HG1 – high grade regional, VHG – very high grade, MG2 – medium grade (subduction) and HG2 – high grade (subduction).

More detailed information about each compositional category at each metamorphic grade is given in Chapter 8. We use a consistent colour coding (e.g. green for category B, mafic igneous rocks), to help you see the relationships between the tables and Figures. Even this simplified exercise of comparing Tables 3.1 and 3.2 with simplified *P-T* diagrams for specific rock types (e.g. mafic, pelitic, etc. (labels A–G in Tables 3.1 and 3.2 and Figure 3.7) and schematic diagrams of grade with *P-T* and tectonic setting (labels VLG, LG, MG, etc.; Figure 3.6) will help you to build an understanding of metamorphic classification. The appendix material in Chapter 8 provides more detail about protolith variation, mineral assemblages, and additional useful reference tables that may be more relevant to the specifics of your field area, as you become more familiar and confident in your understanding.

3.3 Metamorphic Rock Names

The names of metamorphic rock-types that can be given in the field are generally of three kinds:

1. Names based on the presence of specific *metamorphic minerals*. These names should be used whenever the metamorphic minerals can be identified.
2. A metamorphic term appended to a pre-metamorphic rock-type name.
3. A summary of the main physical properties. These are used for rocks in which *minerals are too fine-grained for individual identification.*

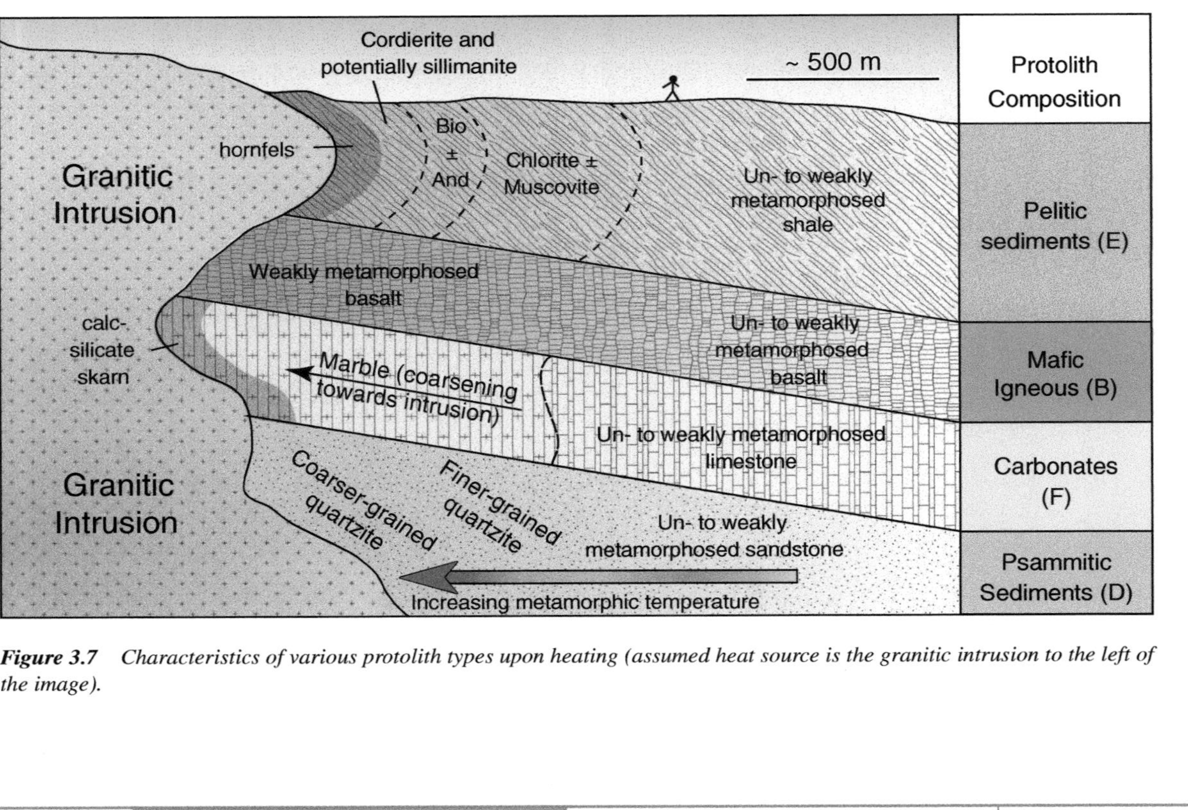

Figure 3.7 *Characteristics of various protolith types upon heating (assumed heat source is the granitic intrusion to the left of the image).*

Other names can be used for special rock types, in particular those of fault-zones and deformation zones (see Table 7.1). Table 3.2 highlights some classic rock types with their classification names for further reference in this section.

3.3.1 Mineral-based rock names

Most metamorphic rocks belong to one of the standard types defined in Table 3.2 (see also Section 8.2.4). *Use a name on this list whenever possible*, but add to it as appropriate (see below). These names have generally accepted meanings in terms of mineral contents, and should not be used in any other way. For example, an amphibolite is a rock which contains plagioclase and hornblende (Table 3.2), so 'amphibolite' should not be used for rocks consisting of an amphibole alone. Additional minerals should be added as *prefixes* to the rock-type name-base, e.g. diopside-bearing marble, garnet amphibolite, garnet micaschist, etc. In the case of micaschists and marbles, the name of the mica or carbonate mineral, respectively, can be added as a prefix to make the name more specific.

Rocks which do not fall into these common categories can be given names by adding a textural type-name (schist, gneiss, hornfels) after the name of the most common mineral or mineral pair (e.g. hornblende schist, chlorite-actinolite schist). If no obvious textural term applies, the suffix '-ite' or the word 'rock' is added to the name of the most common mineral, and other mineral names may precede it (e.g. garnetite, diopside-bearing hornblendite).

3.3.2 Names based on pre-metamorphic type

The most basic of these names are 'meta-' prefixed to a previous rock-type name (e.g. meta-arkose). Whenever possible, *a textural term should be included*, to indicate the nature of the change that has occurred (e.g. *schistose* granule conglomerate, *hornfelsed* micaschist, *gneissic* meta-granite, *flaser* gabbro). The 'meta-' prefix may be dropped, if doing so leaves no ambiguity. The name 'granite gneiss' is intentionally ambiguous, but is justified only where it is really unknown whether deformation has occurred before or after the solidification of the rock. Otherwise an unambiguous name should be used. In the case of a granitic rock which has definitely undergone deformation as a solid rock, the term 'gneissic meta-granite' is more precise. The prefixes ortho- and para- are also sometimes used to refer to gneisses of igneous and sedimentary origin, respectively.

3.3.3 Names of fine-grained metamorphic rocks

The basis of fine rock-type names is either physical: *hornfels, slate, mylonite*, etc.; or compositional: *marble, quartzite, greenstone, serpentine*. Both terms can be used if possible (e.g. mylonitic carbonate). These name-bases can have words added to them that describe additional key sample-scale properties such as colour (e.g. black slate) or visible minerals (e.g. pyritiferous slate). Sometimes overall properties have compositional implications (e.g. spotted hornfels, red slate), although these cannot be specified in detail.

3.4 Reporting Rock Types

In your field study, the reporting of rock types, their classification, cataloguing, and collection, are very important. This is not just to enable a basic mapping of an area, but is also vital for the context of any samples collected for further study. We will consider the way to record rocks with visible minerals (Section 3.4.1) and how to go about recording finer patches within these (Section 3.4.2). A *fine-grained rock-type* may seem much harder to characterise but should be introduced by its name, and then described as fully as possible (See Section 3.4.3).

3.4.1 Moderate- to coarse-grained rocks

A *rock type in which minerals are visible* should be introduced by *(1) its name*, followed by *(2) a list of constituent minerals* in order of decreasing abundance. Each mineral name may also include descriptors such as 'porphyroblastic' or 'idiomorphic'. If mineral proportions are uniform, these should be written next to the mineral names (as percentages). If any mineral is clearly a remnant from an earlier rock type, this too should be indicated.

The rest of the rock-type description should work from the general to the particular, starting with *(3) overall rock properties* (colour, fracture, jointing, schistosity, weathering style, etc.). You should include *(4) sketches of any compositional patchiness or fine-scale banding*, annotated with statements clarifying mineral proportions in the different parts. This should be followed by *(5) sketches of grain textures*, annotated with mineral names. All sketches should show scale, and the orientations of any directional features. Two separate strands of rock-type description lead off from the sketches of textures: mineral details (as below); and fabrics, structural context, etc. These are treated more fully in Chapters 4 and 5 (see also checklist in Section 8.2.5).

Either as annotations to sketches, or separately, there must be *(6) statements about which minerals sometimes occur in mutual contact, and which do not*. It is worth checking that mineral proportions, grain sizes, and mutual contacts have all been recorded fully up to this point in the description. When this is done, there should be (where necessary: see Section 3.1) *(7) brief descriptions of each mineral*, and then *(8) descriptions of any fine-grained portions* (See Section 3.4.2). For this purpose, minerals can be considered to fall into four groups (as in Section 8.2), to be reported in this order:

1. *Common bulk minerals*. Between them, these usually make up most of the rock. They normally belong to the following list: quartz, feldspars, olivines, pyroxenes, amphiboles, micas, carbonates, serpentine minerals, talc, chlorite, and epidote.
2. *Metamorphic minerals*. Though chemically composed of the same components as bulk minerals, they are generally present in smaller amounts, and are sometimes individually diagnostic of particular metamorphic grades or compositional categories (e.g. Table 3.2 and Chapter 8). This group includes: garnets, aluminium silicates, chloritoid, staurolite, cordierite, prehnite, zoisite, and lawsonite. *In marbles*, talc, amphiboles, pyroxenes, and olivines can be considered in this category.
3. *Accessory minerals*. These carry minor chemical elements (boron, zirconium, titanium, etc.) in concentrated form. They include the oxides of iron and titanium, sulphide minerals, sphene (titanite), apatite, beryl, tourmaline, and chromite. Frequently, these exist in a metamorphic rock at the same location as did accessory minerals (rich in the same elements) in the pre-metamorphic state, and a note should be made of any evidence suggesting that the present grains existed before metamorphism.
4. *Minerals of unknown identity* (if any). These should be described as fully as possible, with sketches of habit, cleavage, alteration to other minerals, etc.

3.4.2 Fine-grained patches, pseudomorphs, and mylonite streaks

Portions of fine-grained material may occur within a rock in which mineral grains are otherwise visible. The two general questions to ask in such cases are, 'How did such a fine-grained entity arise?' and 'What is it made of?'. The possibilities are that:

1. Such portions existed as fine-grained entities in the pre-metamorphic state.
2. Grain size has been reduced by chemical processes of metamorphism (pseudomorphing, in a broad sense).
3. Grain size has been reduced by deformation.

Fine-grained chunks, either rounded or angular, having forms like large clasts are usually pre-metamorphic. Often, they were lithic clasts in coarse sediments, bioclasts, or concretions (chert, ironstone, or calcareous) in fine sediments, or xenoliths in igneous rocks. Almost all fine-grained rock-types can occur as lithic clasts or xenoliths. When found in metamorphic rocks, these fine materials may be treated and described as metamorphic rocks in their own right (Section 3.4.3), though then association with the coarser rock types in which they are found should be noted, as should any potential reactions between the two (see also Chapter 6).

If you find many fine-grained chunks that are the size of large crystals and show consistency of composition, size, shape, and distribution through the rock, these may be pseudomorphs (see Section 5.2.2). If they are aggregated, or vary in size, this may reflect the original grains of the mineral that they pseudomorph (e.g. clusters of plagioclase or pyroxene grains in a gabbro, or strings of phenocryst minerals parallel to the walls of a dyke). Pseudomorphed minerals may predate metamorphism or may be early metamorphic minerals that were replaced later in the metamorphic history. Although many pseudomorphs have distinctive angular shapes, this is not essential. Just as minerals, of metamorphic rocks in particular, are commonly ovoid or irregularly polyhedral, so will be pseudomorphs formed from them. Beware of the capacity of some minerals to form porphyroblasts by including large numbers of much smaller grains of other minerals. These can look like pseudomorphs, because of the finer grains within them, but they are not. The smaller grains inside such porphyroblasts are usually more similar in mineral species, size, and shape to the grains of the matrix than is the normal case for grains of a pseudomorph. In many cases, the aggregate of minerals in the pseudomorph are too fine-grained to be seen with the naked eye, potentially leading to the incorrect interpretation that a primary mineral is preserved rather than a suite of minerals that have replaced it (Figure 3.8). Late modifications to shapes may have occurred. Reaction tends to round off the corners of angular pseudomorphs. Deformation may do the same, and may also create an array of strained shapes which could not have accommodated the earlier pseudomorphed mineral. Pseudomorphs may also be zoned. Pseudomorphs are considered in more detail in Section 5.2.2.

A significantly finer-grained matrix than abundant embedded coarser grains or chunks of coarser rock may occur in samples that look like either a metamorphosed clastic sediment or a meta-volcanic rock. Those interpretations might be correct, but similar outcomes can also be the product of deformation: The distinction can sometimes be difficult. Whether the fine material is pre-metamorphic or produced by deformation, the coarser parts are likely to have rounded-off edges, and the matrix might be either felsitic and generally nondescript or dark and slaty. The matrix of quartzo-feldspathic rocks is likely to be slaty or sericitic, whereas the matrix of intermediate or mafic igneous material is likely to be more chloritic. Deformation-induced grain-size reduction may be proved by finding either pieces of an original coarse grain that has been pulled apart parallel to the foliation, or a streak of fine material cutting through coarse grains. It is important to remember that this grain-size reduction may still have been superimposed on a rock that previously had a fine matrix. The distinction between these possibilities is best made by mapping out the distribution of fine and coarse zones, and by a comparison of associated rock types. This should show whether their occurrence is within a stratified depositional sequence (clastic, volcanic, or volcaniclastic) or in a deformation zone. Rocks of deformation zones, particularly cataclasites and mylonites, are the subject of Chapter 7.

3.4.3 Fine-grained rocks

Rocks that cannot be described in terms of individual minerals on account of extremely fine grain size should be described in terms of their physical and simple chemical properties: hardness; colour and lustre on fresh and weathered surfaces; resistance to weathering relative to other specified rock types; degree of visible anisotropy and any cleavage; whether homogeneous,

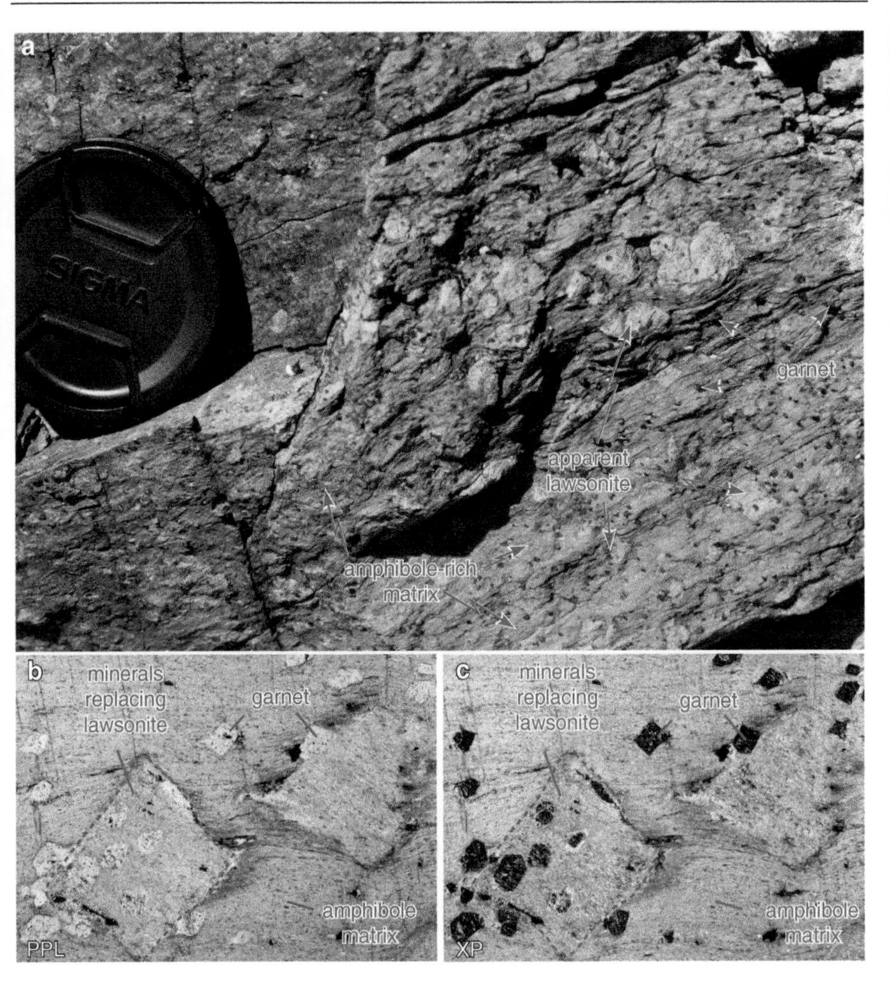

Figure 3.8 *(a) Apparent lawsonite crystals at the outcrop scale. A thin section of this rock shown in plane-polarized (b), and cross-polarized (c) light reveals that these crystals are pseudomorphed by a fine-grained aggregate of minerals (width of image in thin section images is ~4 cm) (photos Mark Caddick).*

patchy, or striped (either parallel or oblique to bedding or banding); whether attacked by hydrochloric acid; and whether gradational in grain size through to coarser rock types, and if so, which other properties share in the gradation. These matters overlap with fabrics and cleavage as treated in Chapter 4.

Table 3.3 lists some fine-grained rock types. Note that, apart from serpentine (or 'serpentinite'), these are either fine-grained before metamorphism, or the pre-metamorphic rock is texturally destroyed by deformation. Fine-grained siliceous rock types are not easily distinguishable from one another, and no suitable and generally recognized common term (roughly the metamorphic equivalent of 'felsite') exists. The term 'greenstone' is not always limited in its usage to fine-grained rocks.

Table 3.3 Some fine-grained rock types (colours indicate protolith see table 3.1).

Hornfelses and Similar Rock Types *Hornfelses*. The process of 'hornfelsing' (turning harder, more brittle and increasingly isotropic) is not restricted to a particular protolith type. *Meta-cherts*. Can be hornfels-like and anywhere between black, hematite-red or bleached. They can occur amongst other metasediments which are not hornfelses. *Hornfels-like siliceous rocks* can sometimes be found amongst other metamorphosed, but not hornfelsed, sediments and lavas. They can be derived from fine-grained felsic igneous (e.g. rhyolitic or aplitic) or silica-rich sedimentary (arkosic or quartzite) protoliths.	**Altered Igneous Rocks** (which may be slaty or isotropic) *Serpentinites* reflect an ultramafic protolith. They may be serpentine-dominated but can contain numerous other minerals (see Table 8.2) *Greenstones* represent a fine-grained mafic or intermediate protolith. The colour is usually that of chlorite or actinolite. Similar brown or grey rocks are possible at Very Low Grade among apparently unmetamorphosed sedimentary *Red-and-green patchy rocks* (no standard name applies) have carbonate-rich patches, hematite-red patches, and epidote-green patches. These can be formed by metamorphism of hydrothermally altered, generally mafic volcanic materials.
Slates (varieties formed from fine-grained sedimentary rock) *Soft slates or phyllites* can consist almost entirely of sheet-silicate minerals (sericites and chlorites). Typically from pelitic protoliths. *Marly slates* may contain detectable calcite co-existing with sheet-silicates. *Siliceous slates* consist of a substantial amount of quartz (and sometimes feldspar) co-existing with sheet-silicates	**Mylonites and Other Faulted Rocks** (see also Chapter 7) *Fault gouge* (incohesive), *pseudotachylyte* (glassy), *cataclasite* (isotropic) and *mylonite* (finely banded, sometimes slaty) can form from all kinds of rock-types.

3.5 Compositional Category and Metamorphic Grade

Indications of previous rock type may be shown by relict sedimentary or igneous structures, and by grains of minerals that have not been destroyed by metamorphism. This information should be checked against, and incorporated with, information available from mineral assemblages, to give an idea of the nature and origin of the rock unit. This section helps in that regard, providing a framework for inferring classes of pre-metamorphic compositional rock type and of metamorphic grade from the present minerals.

3.5.1 Grade of metamorphism

In describing a rock mass, transitions from one set of minerals to another should be noted, and if possible mapped out. The question always arises, 'Does this change in minerals represent a change in bulk chemical compositions, or a change in the grade of metamorphism, i.e. an *isograd*. Therefore, deductions about changes in composition and about grade must be tackled together, with both requiring two levels of observation. Variations of rock type *within an outcrop* can be described and explained either in pre-metamorphic compositional terms, or as changes in metamorphic grade on a local scale. These relationships are vital to the geological synthesis of the area, because very local changes in metamorphic grade are diagnostic of particular drivers of metamorphism, such as fluid and melt input. The other level is that of *external reference*, comparing and categorizing rocks in terms of an external, generally understandable scheme. Section 8.1 provides such a scheme, defining compositional classes for common metamorphic rocks and giving information for classifying their grades. This is simplified in Tables 3.1 and 3.2. It defines compositional classes for common metamorphic rocks, and gives information for classifying their grades. Determinations of absolute pressures and temperatures of metamorphism are beyond the scope of non-specialist description, but recognition of key mineral associations can give some initial indication as to whether they are broadly *low grade (LG)*, *medium grade (MG)*, etc., as summarized in Table 3.1. These are broad divisions. If some rocks in an area occur partially as *low grade* and partially as *medium grade*, they can be described as transitional between the two. In this case, the differences might reflect slight differences in metamorphic pressure or temperature across the area, differences in fluid interaction with rocks within the area, or subtle changes in protolith composition across the area. In some cases, identification of one of the polymorphs of Al_2SiO_5 may be sufficient to determine that a rock experienced high-grade metamorphism, for example if sillimanite is present, or that it experienced medium-grade metamorphism at low pressure, if andalusite is present (Figure 3.9).

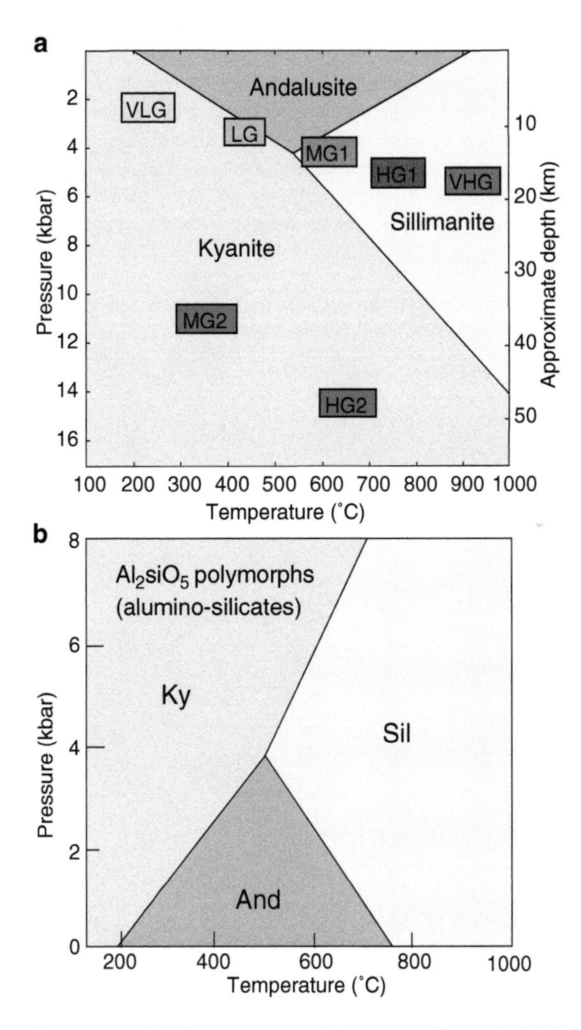

Figure 3.9 *Stabilities of the Al_2SiO_5 polymorphs (aluminium silicates). (a) Labelled on the P-T diagram, with approximate indications of the grade indicators used throughout this book. (b) A common alternative way in which you will see this represented (with P reversed).*

It is important to recall, however, that each of the polymorphs is stable over a large swath of *P-T* space, so this is not an extremely sensitive tool for grade determination. Furthermore, the polymorphs can often persist metastably in a rock that has exceeded the nominal limit of their stability.

3.5.2 Triangular (compatibility) diagrams and grade of metamorphism

In more accurately determining how rocks change upon increasing metamorphic grade, it is useful to understand what reactions between minerals are possible and at what range of conditions those reactions might be expected to occur. This is commonly done using triangular 'compatibility diagrams' (e.g. Figure 3.10b), the differences between which reveal possible mineral reactions (Figure 3.10a). For example, in pelitic rocks that also contain quartz and muscovite, and have an available hydrous fluid, the association chlorite + staurolite is generally not stable above about 600 °C, instead forming biotite + kyanite (the line between fields 'iv' and 'v' in Figure 3.10a). Thus, recognition of coexisting chlorite + staurolite places an approximate *upper limit* on the temperature of metamorphism. By the same logic, recognition of coexisting biotite + kyanite places a *lower limit*. The mineral chloritoid is shown to break down to chlorite + garnet + staurolite along the equilibria between the fields labelled 'ii' and 'iii', thus the presence of chloritoid in a rock implies that it did not exceed metamorphic temperatures of approximately 575 °C. However, the co-stability of both chloritoid *and* kyanite is lost at lower temperature, between fields 'i' and 'ii'. Figure 3.10a thus represents a calculation of the

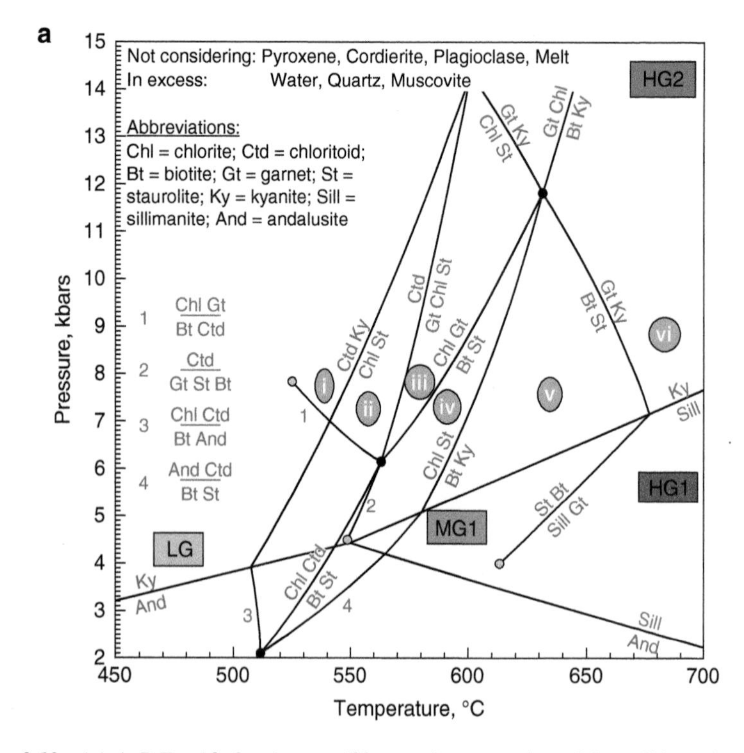

Figure 3.10 *(a) A P-T grid showing possible reactions experienced by pelitic rocks and the approximate grade indicator (LG, MG1, etc.), and (b) accompanying AFM compatibility diagrams. See text for more information. Calculated with the program THERMOCALC (see https://hpxeosandthermocalc.org).*

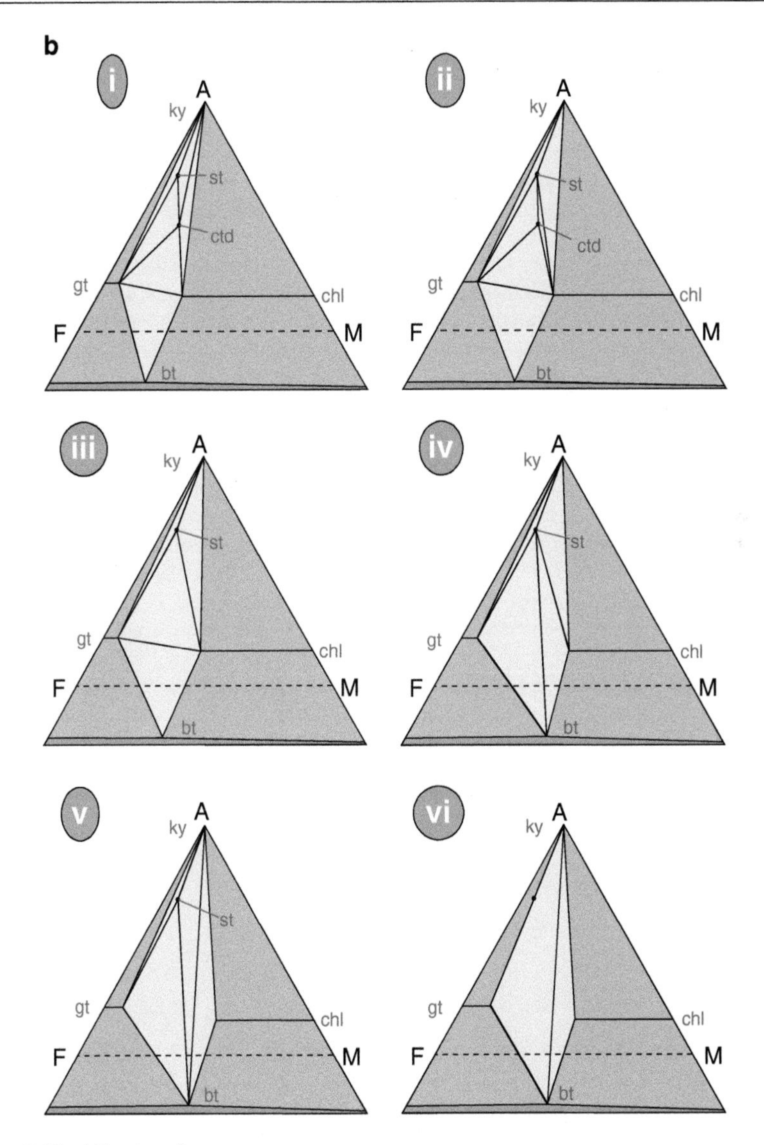

Figure 3.10 (*Continued*)

terminal stabilities of mineral associations (always assuming the presence of quartz, muscovite, and water). It is a significant simplification of natural pelitic rocks, with the calculation only considering K_2O, FeO, MgO, Al_2O_3, SiO_2, and H_2O and being dependent on the validity of the underlying thermodynamic properties that went into its calculation. It is, however, a useful guide.

The specific minerals that *will* occur in a rock are shown in the triangles in Figure 3.10b, because the bulk composition of the rock also dictates the mineralogy and Figure 3.10a only shows the extent of mineral associations that are *possible* as a function of *P* and *T*. The AFM triangles in Figure 3.10b address that, allowing you to see the minerals that will occur for a given rock composition at the approximate *P-T* conditions of each triangle. In these, the A, F, and M apexes relate to the Al_2O_3, FeO, and MgO contents of the bulk rock and the minerals within it. Note that the 'A' apex also accounts for the K_2O content, such that the diagram is 'projected' from muscovite (demanding that muscovite coexists with all of the mineral assemblages in the triangles). A low aluminium pelite at the approximate *P-T* conditions of triangle 'iii' will contain chlorite + biotite if it is MgO rich (i.e., if its composition plots in the green field on the dashed F–M line, closer to M than F); will contain chlorite + biotite + garnet if it has a slightly higher FeO:MgO ratio than this (if its composition plots in the paler green field with chlorite + biotite + garnet at its corners); and will contain just garnet and biotite if it has an even higher FeO:MgO ratio. In every case, it will have equilibrated in the presence of quartz, muscovite, and water. It should be clear that more aluminous bulk compositions (i.e. closer to the 'A' apex) are required to form assemblages with kyanite or staurolite at these low-temperature conditions. The simplified reaction chlorite + garnet = biotite + staurolite, shown in Figure 3.10a, implies that at higher temperature (e.g. in AFM triangle 'iv'), staurolite becomes stable even for rocks of only moderate aluminium content, for example in the association garnet + biotite + staurolite. This is relatively uncommon, but field evidence for it can be found (Figure 3.11).

The AFM triangles at different *P-T* conditions are useful guides to the mineral assemblages that are likely in pelitic rocks. Alternative plotting schemes are more appropriate for mafic rocks and calc-silicates, with the positions of minerals in 'ACF' and 'SiO_2–CaO–MgO' plotted in Figure 3.12. It may be useful to make copies of these in your notebook, so that you can draw in sub-triangles for mineral assemblages in your field area, as shown in Figure 3.13. These can help you to determine how variations in bulk-rock composition may be controlling differences in mineral assemblages across an outcrop and how reactions between minerals might occur. More detail on the progressive reactions that occur in each of these systems and how they change the topology of tie-lines within the three triangles shown in Figure 3.12 is beyond the scope of this text, but we recommend that you follow up the detail on metamorphic reactions in more advanced texts (such as Spear (1993) and Best (2003), see further reading suggestions).

Figure 3.11 *The Barrovian sequence can be deceptive! This staurolite, garnet, and muscovite schist is so aluminous that staurolite grew at low-to-moderate grade, before biotite stability (photo Mark Caddick, from the western Himalaya of NW India).*

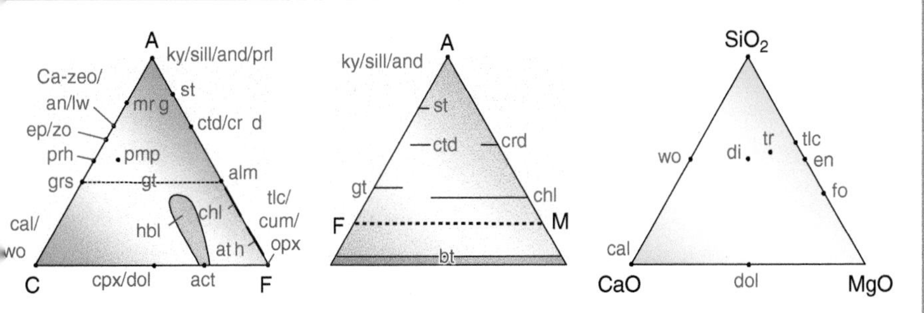

Figure 3.12 The compositions of key minerals in mafic, pelitic, and calc-silicate rocks. Black dots are mineral compositions, black lines and shaded fields are mineral compositional ranges. In the ACF diagram for mafic rocks, minerals are plotted according to a modification of Eskola's classic work, in which A = $AlO_{3/2}$, C = CaO and F = FeO + MgO (see Spear, 1993, for more details). AFM is projected from quartz, muscovite and H_2O, with A = Al_2O_3–3 K_2O, F = FeO and M = MgO (following Thompson, 1957). Biotite plots below the F–M join due to this projection. Note that an alternative projection is required at higher-grades, where K-feldspar replaces muscovite. In the third diagram, phase compositions are plotted as a function of their direct oxide proportions. Mineral abbreviations are as follows: ky = kyanite, sill = sillimanite, and = andalusite, st = staurolite, ctd = chloritoid, crd = cordierite, gt = garnet (alm and grs = almandine and grossular end-members), chl = chlorite, bt = biotite, cal = calcite, dol = dolomite, wo = wollastonite, cpx = clinopyroxene (di = diopside end-member), hbl = hornblende (tr and act = tremolite and actinolite en-members), cum = cummingtonite, tlc = talc, opx = orthopyroxene (en = enstatite end-member), fo = forsterite, prh = prehnite, pmp = pumpellyite, ep = epidote, zo = zoisite, an = anorthite, lw = lawsonite, Ca-zeo = laumontite, heulandite, stilbite, or wairakite.

3.5.3 Major compositional categories

Section 8.1 covers all of the commonest metamorphic rocks formed from massive pre-metamorphic parents. It excludes most reaction-zones, metasomatic rocks, hydrothermal deposits and metamorphosed soils and subsoils.

We divide meta-igneous rocks into ULTRAMAFIC (A), MAFIC (SOMETIMES TERMED 'BASIC') and INTERMEDIATE (B), and FELSIC (SOMETIMES CALLED 'ACID') (C) categories. Metasediments are divided into QUARTZITES (D), SEMI-PELITE and PELITE (E), CALCAREOUS (F), and META-MARLS and CALC-SILICATES (G). *Pelites proper* are a sub-class (E+). A number of transitional types are possible between the categories used here, but the information should be sufficient to identify their pre-metamorphic type and their metamorphic grade. Large clasts, xenoliths, intrusions, or cumulate layers of one composition may exist within a host rock of a different category, for example in a sequence of sediments that were intruded by mafic dykes prior to metamorphism.

A systematic overview of the genesis and mineralogy of these compositional categories is given in Chapter 8, hopefully minimizing the need for cross-referencing with this section. The following sections of this chapter, however, briefly outline some of the key characteristics of the categories upon metamorphism. Where specific grades are discussed, we use the colour-coded scheme that appeared in Figure 3.6a and is discussed further in Section 8.1. Summary Table 8.2 shows the most common minerals of some of the compositional categories at various grades. These minerals are not necessarily grade indicators, which are discussed further in Section 8.1. These grade indicators may be *individual minerals*, or a *combination* of two or more minerals, which may be only a part of the complete *assemblage*. The word combination is used here when a grouping of minerals indicates grade, but does not constitute the whole rock. Table 3.2 identifies the diagnostic minerals of specific rock types, highlighting the compositional category that they are most closely affiliated with.

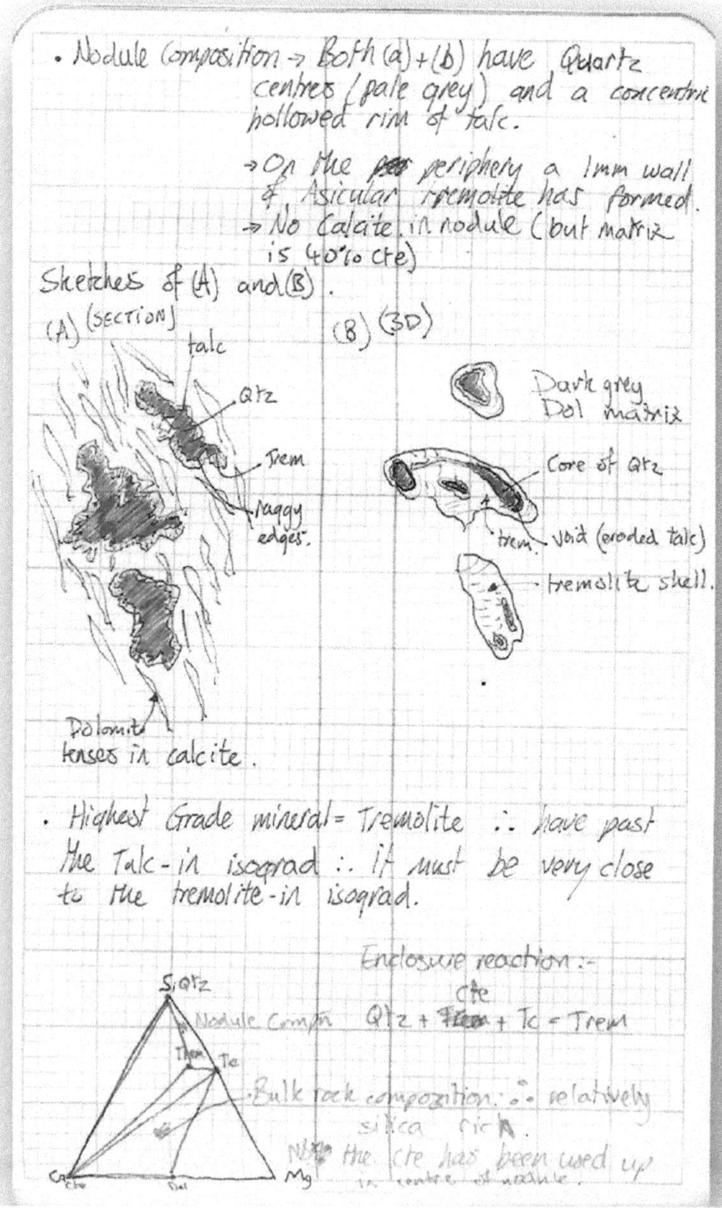

• Nodule Composition → Both (a) + (b) have Quartz centres (pale grey) and a concentric hollowed rim of talc.
→ On the periphery a 1mm wall of Asicular tremolite has formed.
→ No Calcite in nodule (but matrix is 40% cte)

Sketches of (A) and (B).

(A) (SECTION) (B) (3D)

talc
Qtz
Trem
raggy edges.
Dolomite lenses in calcite.
Dark grey Dol matrix
Core of Qtz
trem.
void (eroded talc)
tremolite shell.

• Highest Grade mineral = Tremolite ∴ have past the Talc-in isograd ∴ it must be very close to the tremolite-in isograd.

Enclosure reaction:-
cte
Qtz + Tc = Trem

Si, Qtz
Nodule Comp.
Trem
Tc
Bulk rock composition: ∴ relatively silica rich.
Note the cte has been used up in centre of nodule.
Ca
Cte
Dol
Mg

Figure 3.13 *Notebook example where a triangular diagram has been drawn to help understand mineral assemblages seen in the outcrop and how they relate to zones of different composition (courtesy of Nick Timms).*

3.5.4 Mafic (also called basic) and intermediate igneous (compositional category B)

Igneous plagioclases metamorphose to calc-aluminous rock portions, and large igneous mafic minerals to mafic portions, with limited interaction other than in some spectacular instances (as shown in Figure 3.14). Any quartz typically remains stable upon metamorphism, as does biotite except at VLG or VHG. Therefore:

1. The ratios of MAFIC: CALC-ALUMINOUS: QUARTZ: BIOTITE generally correspond relatively well to those of the precursor igneous rock.
2. Igneous structures generally remain visible unless deformation has occurred.
3. Coarse plutonic textures remain as pseudomorphs or augen. In this case, careful characterization of composition and structure can identify the igneous parent.
4. Rocks comprised of homogeneously intermixed calc-aluminous and mafic minerals derive *either* from fine-grained igneous rocks, *or* from coarse-grained igneous rocks by deformation, *or* from rapidly resedimented igneous material. These are all likely to be banded in some way.

3.5.5 Pelitic rocks (compositional category E+)

These rocks have traditionally been characterized by the first occurrence of indicator minerals including biotite, garnet, staurolite, kyanite, or andalusite in line with the classic Barrovian Zones (e.g. Figure 3.15, see also Figures 3.1 and 1.3). However, none of these minerals is a reliable indicator of absolute conditions of metamorphism if rock compositions are highly variable. The AFM triangles shown in Figure 3.10 demonstrate that, for example, very aluminous pelitic rocks can contain kyanite or sillimanite before the initial stability of biotite. This is contrary to the classic ordering of the Barrovian sequence. Mineral indicators may be used to define zones locally if it can be demonstrated *in the field* that they give a consistent zonation. Even then, it is possible that they may not allow very close reference to universal grade categories (as in Section 8.1).

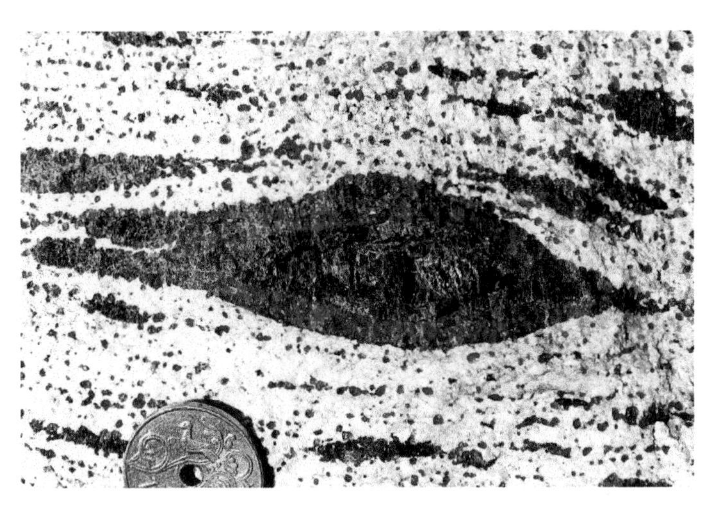

Figure 3.14 *Reactions between contrasting compositional portions of a rock can be spectacular! Impurities in anorthosites have reacted with their host upon metamorphism to yield a concentric pattern of (from the core of the structure) orthopyroxene, clinopyroxene, garnet, and anorthite (plagioclase), Holsnøy, Bergen Arcs, Norway (courtesy of Chris Clark).*

Figure 3.15 *Representative rocks of the Barrovian sequence, and their approximate grade indicator. Grain size is so small in the lower-grade samples that individual grains are difficult to discern at this scale (photos Mark Caddick).*

In *VLG* pelites, both absolute metamorphic conditions and rock compositions can produce variability of mineral assemblages, but as several of the minerals are 'sericites', detailed distinctions are generally not possible in the field.

If carbonate minerals occur at *VLG* or *LG* in rocks which would otherwise be classified as pelitic, such rocks should be classified as 'meta-marls' (compositional category G), because they may give rise to quite distinct calc-silicate rock-types at *MG*, *HG*, and *VHG*.

3.5.6 Marbles and calc-silicates (compositional categories F and G) and fluids

In silicate rocks, prograde metamorphism proceeds by dehydration. If retrograde reactions occur, they rehydrate higher grade minerals, producing minerals which had previously occurred in the prograde metamorphism.

In marbles, prograde metamorphism proceeds by decarbonation. Retrograde hydration reactions produce hydroxides and hydrous silicates which did not occur during the prograde evolution. Therefore, some late *LG* minerals can be good evidence of earlier higher grade metamorphism.

Marls contain both silicate and carbonate portions. The silicates dehydrate substantially through *VLG* and *LG* conditions, while decarbonation reactions occur mainly at *MG* and *HG*. Through *MG* in particular, carbon dioxide and water dilute each other. This tends to reduce the temperatures at which minerals first appear, and can cause both sides of an expected reaction, such as:

$$\text{anorthite} + \text{calcite} + \text{quartz} = \text{grossular} + \text{carbon dioxide}$$

to persist through an interval of metamorphic grade, as determined by other rock types. At higher grades the distinctions between different compositions (e.g. carbonate vs silicate rich) can be quite subtle (e.g. Figure 3.16).

For practical purposes, therefore:

1. Grades indicated by particular minerals are not the same in pure marbles, impure marbles, and in silicate rocks.
2. It is important to record whether a carbonate mineral exists in each calc-silicate rock type.
3. The most informative calcareous assemblages are those with the most minerals.
4. A scheme for determining grade of metamorphism, such as is given in Section 8.1, will generally not apply to calcareous rocks being metamorphosed a second or subsequent time.

Figure 3.16 *Reactions between contrasting compositional portions of a rock can also be subtle! A carbonate-rich lens (greenish region in the top third of the photograph) in a pelitic sediment is now difficult to identify in this rock that experienced migmatization at ~800 °C, from SW Virginia, USA (photo Mark Caddick).*

5. Minerals found near contacts (or in cross-cutting veins) of marbles and of silicate rocks may occur at anomalous grades as a result of carbonate and hydrous fluids diluting each other.
6. Generalizations about grade can only be approximate for these rocks.

3.5.7 Mafic calc-silicate composition overlap (categories B and G)

The reactions in calc-silicates that incorporate the calcium and magnesium from carbonates into silicates (such as plagioclase, actinolite, or diopside) can turn dolomitic marls into mineral assemblages more typically associated with meta-igneous rocks. However, these are compositional flukes within the considerable variability of mixed source clastic/carbonate sediments. Calc-silicate and meta-igneous rock sequences should be distinguishable by their different modes of compositional variation:

1. Uniform meta-igneous types display a degree of compositional control which is only explicable in terms of magma genesis, not sedimentation. They are meta-igneous.
2. Rocks which vary only in the proportion of calc-aluminous to mafic material, and may contain some quartz and biotite, show a control of compositional variation that can be explained in terms of igneous mineral segregation, not sedimentation. They are meta-igneous.
3. Sequences which range from highly calcic (rich in carbonate or diopside), through meta-igneous to pelitic or semi-pelitic (often rich in quartz and biotite) show the kind of variation of meta-marls. They are meta-sedimentary calc-silicates (compositional category G).
4. More restricted calc-silicate compositions can develop by metasomatism or in reaction-zones (Chapter 6).

The features mentioned in (1) and (2) may be retained in rapidly re-sedimented igneous material or in volcaniclastic rocks, in which sedimentary structures and bedding may be visible.

3.5.8 The loss of white mica + quartz: 'high grade'

In most cases, the prograde breakdown of *white mica + quartz*, either by partial melting or by reaction to *K-feldspar + Al-silicate*, provides a useful definition of the transition between *MG* and *HG*. However, in calc-silicates (compositional category G), these minerals breakdown by reaction with calcite at (rather than above) *MG*. This occurs through the reaction:

$$\text{muscovite} + \text{calcite} + \text{quartz} = \text{K-feldspar} + \text{anorthite} + CO_2 + H_2O$$

For pelitic protoliths at pressures above ~5 kbars, the loss of muscovite may occur at higher temperature than the initial generation of melt if free fluid is available at the temperature of the solidus, shown in Figure 3.17. It is, however, likely that melting induced by the dehydration of muscovite represents the first point at which a considerable melt fraction is present in most pelitic rocks, given that fluid availability is likely to be very limited at the solidus. At higher temperature still, biotite breakdown will form some combination of garnet, orthopyroxene and/or cordierite, generating additional melt. Although these reactions are drawn as sharp (univariant) lines in Figure 3.17, biotite loss in particular can occur over an interval of several tens of degrees.

Coexistence of a calcium-free aluminous mineral with K-feldspar is diagnostic of *HG* or *VHG*. At *VHG* conditions, mineral assemblages have typically been thoroughly dehydrated and orthopyroxene-bearing, granulite-facies assemblages dominate.

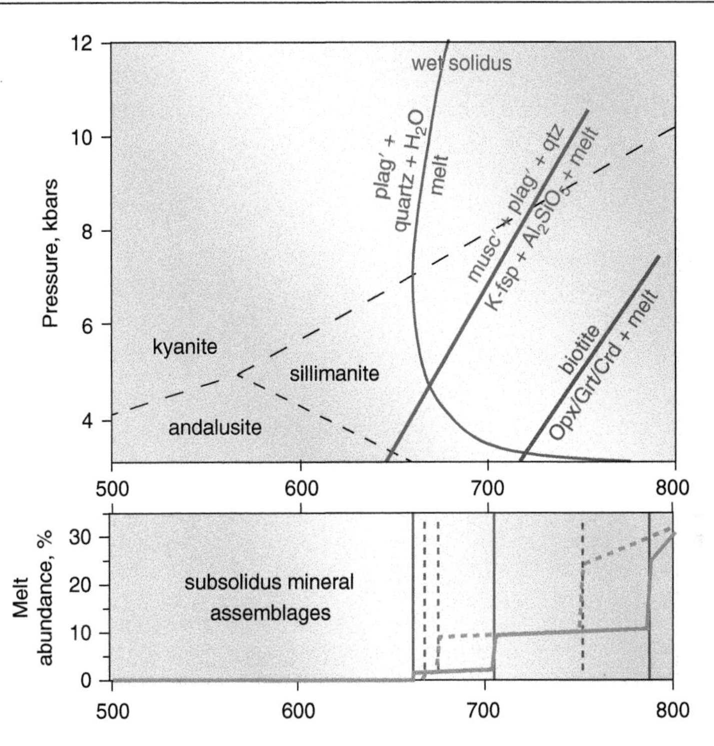

Figure 3.17 *A schematic PT grid showing the three major reactions associated with the generation of melt in pelitic rocks at HG and VHG conditions. The stability of Al_2SiO_5 polymorphs is shown for reference. 'Plag' and 'musc' refer to plagioclase and muscovite, respectively. The lower panel tracks the amount of melt (orange lines) as a function of temperature as the three reactions are crossed at example pressures of 5 kbars (dashed lines) and 7 kbars (solid lines).*

UNDERSTANDING TEXTURES AND FABRICS 1: BANDING, CLEAVAGE, SCHISTOSITY, AND LINEATIONS

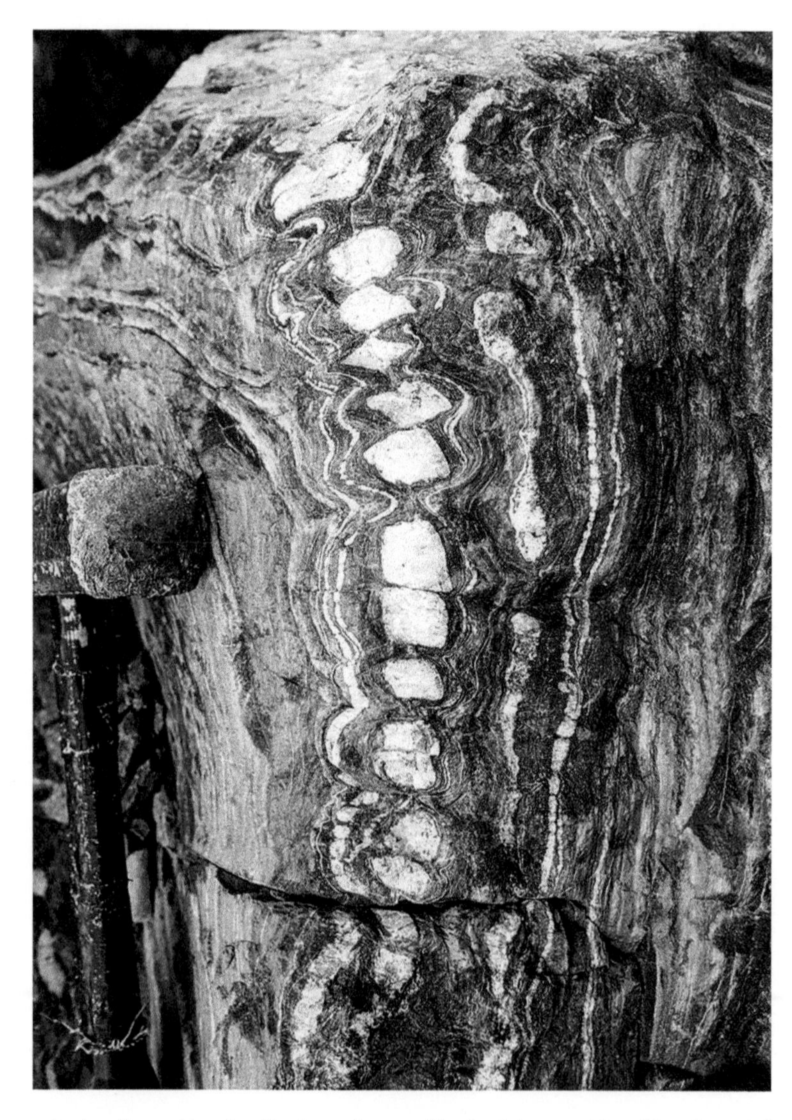

Spectacular boudinaged bands, affectionately named by the Oslo team 'dental boudinage', from the Seve Nappe Complex in Northern Sweden (Sarek) (photo courtesy of Hans Jørgen).

4

UNDERSTANDING TEXTURES AND FABRICS 1: BANDING, CLEAVAGE, SCHISTOSITY, AND LINEATIONS

Most of the rocks that we encounter in the field show some type of compositional banding. This can be formed or enhanced by metamorphic processes that acted on the rock or may just represent initial compositional variations of the protolith. Additionally, a striking feature of many metamorphic rocks is the development of secondary planar features and lineations, imparted as new fabrics during heating and deformation. Very prominent examples of this include the formation of metamorphic gneissose and migmatitic banding, the latter formed upon internal generation of melt (e.g. Figure 4.1). In this chapter we will discuss banding types and scales, and consider the common types of planar features and lineations found in metamorphic rocks. We follow this in Chapter 5 by concentrating on more 'isolated' features that are important in understanding the evolution of metamorphic rocks. Before this, however, we need to introduce the terminology used in these chapters and define some of the implications of this terminology for the resulting textures and fabrics.

4.1 General Terminology

The words fabric, structure, banding, and texture relate to geometric patterns produced by the relative shapes and alignments of mineral grains in a rock. These can impart planar or linear anisotropies that can be manifested at scales from μm-scale grain-aggregates, to bulk properties in hand specimens, to outcrop or regional-scale shear zones (discussed separately in Chapter 7). Several contradictory conventions concern the usage of these terms for metamorphic rocks (and in rocks in general). To help clarify their use within this book, we use the following definitions:

- ***Structure*** refers to disposition, in three dimensions, of portions of a rock that are larger than individual grains (e.g. bands, spherulites, boudins, beds, etc.). In this sense, ***banding*** within a rock typically refers to repeating layers of different composition, generally visible with the naked eye. This banding can be inherited (as in the case of many meta-sediments, which preserve contrasting compositions of different beds) or may have been formed during metamorphism (as is the case in rocks that have experienced partial melting).
- ***Texture*** refers to shapes produced by grain outlines, taking several adjacent grains together (granulose, ophitic, etc.). Whether this is visible in the field will depend on grain size, but texture is often defined at the thin-section scale. Most textural elements involve the juxtaposition of individual crystals with their neighbouring grains, and these are described more fully in Chapter 5.
- ***Fabric*** refers to directionality imparted on the whole rock by the sum of all preferred orientations of elements within it (i.e. by the smallest scale structure, the texture, or both). An example is that if all muscovite grains in a rock are aligned, their *texture* imparts a sample or outcrop scale *fabric* in which some orientations of rock are weaker than others.

Structure is used in a variety of ways and at numerous contrasting scales in geology, for example referring to anything from cm-scale ripple cross-lamination up to the lithospheric structure. Furthermore, a single folded sedimentary bed can be thought of as a juxtaposition of a sedimentary

The Field Description of Metamorphic Rocks, Second Edition. Dougal Jerram and Mark Caddick.

Figure 4.1 *Banded rocks of the Greenland Banded orthogneiss in contact with amphibolite, both heterogeneous and deformed. Contact zone contains large garnet crystals (brown-red minerals left of hammer) (photo courtesy of Susanne Schmid)*

structure (the bed) and a deformational structure (the fold), with no requirement that the two structures are of the same scale, or connotation that they formed by similar processes or at the same time (they didn't). When used in this chapter, 'structure' refers to 'overall structure' at a scale of a hand-specimen/immediate outcrop, or to 'tectonic structure' (or tectonic structures) at any larger scales. Microstructure is generally not defined in the field, being reserved for microscopic and sub-microscopic features. Fabrics are of particular importance in this chapter as they deal with planar or linear features that are imparted in the rock during deformation and are generally readily observed at the hand-specimen to outcrop scale. Textural descriptions and examples are dealt with in detail in the next chapter (5).

If any structure can be identified within metamorphic rocks, it will be marked by variations in the proportions of physical components (e.g. grain shapes, orientations, and sizes) or in chemical compositions (e.g. banding). It is therefore important to have some background understanding of the possible origins of differences in composition and of the means by which these may be destroyed.

4.1.1 Causes of compositional complexity

1. Pre-metamorphic complexity, created by sedimentary, igneous, or diagenetic processes. This can exist at all length-scales. Most compositional variation is of this type.
2. Veining, hydrothermal action and reaction zones (see Chapter 5 for more details). These be usefully split into four sub-groups:
   a. Local veining (segregation veining). Development of fluid-filled cracks or cavities, and their infilling (either as they open or after opening) with material from the local host rock (e.g. Figure 4.2a). This may be by precipitation from an aqueous solution or by crystallisation of partial melt. Such processes are limited to distances of local interstitial fluid flow and produce veins which are usually millimetres or centimetres in width, although widths up to tens of metres are possible.

b. Long-distance veins/intrusions (depositional veins). Masses of externally derived material, introduced and deposited along channels through the rock, as dykes, veins, or pegmatites. These are usually 10 cm–10 m wide, though they can reach almost any larger size as composites of multiple intrusions. Veins may form by a mixture of processes (a) and (b).
c. Metasomatic zones. Successive zones of alteration of a host rock, caused by passage of extraneous fluids and chemical interaction between the fluid and the host. They can be variable in the degree of alteration attained and thicknesses of rock affected, according in part to the pattern of rock permeability (interstitial or fracture) and the reactivity of the fluid–rock combination.
d. Reaction zones. These are new rock compositions that are formed at the interface between previous chemical heterogeneities such as contrasting beds, felsic dykes in carbonates, or mafic enclaves in felsic lithologies. They are generally limited to diffusion distances (metres at most) from contacts.

3. Pressure solution and pressure melting (e.g. Figure 4.2b and c). Broadly speaking, these are processes of stress-induced, deformation-accommodated, metamorphic segregation on a mm to

Figure 4.2 Examples of compositional complexities. (a) Small en-echelon veins, Norway; (b) stylolites within a calcschist, Namibia; (c) complex banding in a contact zone between anorthosite and orthogneiss, showing banded orthogneiss containing boudinaged layers of anorthosite. Archean, Kapisillit mapsheet of western Greenland (photo (a) and (b) Dougal Jerram, photo (c) courtesy of Susanne Schmid).

dm scale (with or without input or removal of outside material). They result in compositional domains with orientations that can be related to the orientations of stress and strain. These processes are collectively known as 'pressure solution' to many geologists, although reprecipitation of the dissolved material can be equally important and is arguably more readily interpreted at the outcrop scale. The diffusion paths involved are typically longer than those around individual grains during either diagenetic compaction or the deformation of a monomineralic rock.

a. *Segregation triggered by local stress variations*. This occurs in rocks having heterogeneous ductilities. When strained, these can develop higher and lower stress areas, leading to subtle variations in the chemical potentials of rock-forming components. Segregation is caused by material *solution*, *transport*, and *re-deposition*, giving a pattern which corresponds to that of the stress variations. This may be an individual *pressure shadow* or *local crenulation*, adjacent to a stronger object such as a garnet porphyroblast, a spatially regular *array of crenulations*, or an *increased segregation of minerals between existing competent and incompetent bands*.
b. *Segregation triggered by chemical reactions, but spatially modified by stress variations around the reaction sites*. This general scenario involves diagenetic or metamorphic dissolution of several minerals with precipitation of new phases (sometimes referred to as *incongruent pressure solution*). This is most significant at Very Low Grade (VLG) and is most common in relatively homogenous sandstones and carbonates. It produces spaced, highly directional, anastomosing compositional stripes, which do not need to conform to any obvious pattern of material ductilities. Stylolites (Figure 4.2b) may be formed by the same processes at lower deviatoric stress.
c. *Dissolution*. Stress variations within a rock can generate sites that will experience preferential dissolution when changes in pressure or temperature either increase the solubility of a mineral or increase the abundance of fluids by mineral dehydration. Upon subsequent removal of the solution, the process corresponds to pressure solution in a stricter sense than does diffusive mass transfer and deposition. Geometrically, this may not be distinguishable from (a) or (b).
d. *Pressure melting*. This occurs when local stress variations lead to the preferential melting of parts of a rock and transfer of either that entire portion into silicate melt (congruent melting) or the generation of both melt and a new mineral phase (incongruent melting). If melt is crystallised locally, this results in segregation (see descriptions of migmatites in Section 4.4.5 for more information on melt-induced segregation). Melt can also be removed from the rock that it was generated in, typically depleting that body in water and felsic components.

4. *Deformation* can enhance the previously mentioned processes of segregation, and increase geometric complexity. It can increase the reactivity of mineral grains, the permeability of the rock, and diffusivities within it. It can bring together masses of rock and extend their area of mutual contact, potentially increasing the potential for reactions between them. Tectonic repetition by thrusting, other faulting, or tight folding can increase the physical complexity of any pattern of compositional variability.

4.1.2 Destruction of compositional complexity

The processes of chemical equilibration tend to convert different parts of a rock body towards a common mineral assemblage and to bring each mineral towards a uniform composition, but they generally do not equalize the proportions of minerals in different rocks. These processes are thus relatively ineffective at destroying compositional heterogeneities at scales much larger than single grains: once a compositional distinction (such as bedding between two distinctively different sediments or dykes in a host rock) exists it will generally be preserved, even if in a very modified form. Cataclasis (see Chapter 7), may destroy the physical ordering of constituent parts of a rock, but short of melting, there are only two processes that can dramatically modify or even destroy compositionally distinct entities.

1. *Development of reaction zones* between rock masses of different composition is driven by chemical potential gradients across the interface that lead to dissolution, transport, and new mineral growth. This can produce wholesale replacement of the borders of one or both rock masses by new minerals with their own textures. If a body is sufficiently small, it may be entirely replaced by its reaction zone and so made compositionally more similar to its host rock (e.g. in Figure 3.16).
2. *Overgrowth by individual mineral grains*. Crystal growth normally destroys many of the details which existed at a finer scale. So, the growth of large grains at higher grades of metamorphism usually destroys the textures of finer grains developed at lower grade. Because different compositional patches in a rock are usually defined by aggregates of grains, these patches become unrecognizable if new crystals grow larger than them. The effect of strain is more important in this than any general increase in grain-size with grade. As soon as an aggregate patch is thinned during deformation to the point of being no wider than the individual grains of which it is composed, it will begin to break up into a string of isolated grains. These will be separated if further strain occurs, and, if the strain is sufficient, the grains of the string will become so widely spaced that the result is not distinguishable from a deformed homogeneous rock.

4.2 Rocks without a Metamorphic Directional Fabric

In many metamorphic rocks, large-scale structure is pre-metamorphic, though new grain textures may sometimes be visible. In this case, any directionality or banding reflects igneous, sedimentary, or compaction processes, and its description can follow naturally from the description of rock composition and banding (see Section 4.3). The extent to which a metamorphic textural history may be discerned can be very limited in this case, as is the scope for textural correlations between different rocks. These 'pre-metamorphic' features are all essentially geometric, so their field description can be predominantly in the form of sketches. Any required text, such as mineral names, can normally be annotated directly onto a sketch, which should always include both a scale bar and the orientation of the diagram. It is important to note that description of rocks lacking a discernible metamorphic fabric (at least, in the outcrop that you are investigating) may be useful at a later date, as it might help to unravel the evolution of similar lithologies that experienced more intense metamorphism or deformation, by allowing you to more clearly define pre-existing features.

Some types of metamorphic rocks, such as those formed by heating upon contact with an igneous intrusion (see Section 1.3.3 and Chapter 6), may evolve in the absence of a strongly dominant stress direction. In this case, a well-developed metamorphic mineral assemblage may form without development of a metamorphic directional fabric. Similarly, some metamorphic mineral growth can occur late in the history of deformed metamorphic rocks, showing either no preferred orientation or an inherited orientation if crystallisation outlasted deformation. These types of observation will be dealt with in more detail in Chapter 5.

4.3 Banding

'Banded rocks' consist of a stack of compositionally definable planar layers. All exposures of these rocks will yield a banded appearance, unless the outcrop surface is completely parallel with the layering, and the apparent thickness of each layer is a function of both its actual thickness and its orientation relative to the examined outcrop surface. If such rocks are encountered, an attempt should be made to characterize the banding pattern. The rock types constituting each of the different bands can be described in terms of their mineral contents and textures, paying particular attention to the growth of different metamorphic minerals at the interface between bands. The modes of occurrence of compositional bands are varied, with Table 4.1 suggesting how different types of bands may be initially distinguished for purposes of outcrop description. See also Figures 4.1 and 4.2c.

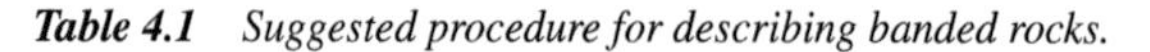

Table 4.1 *Suggested procedure for describing banded rocks.*

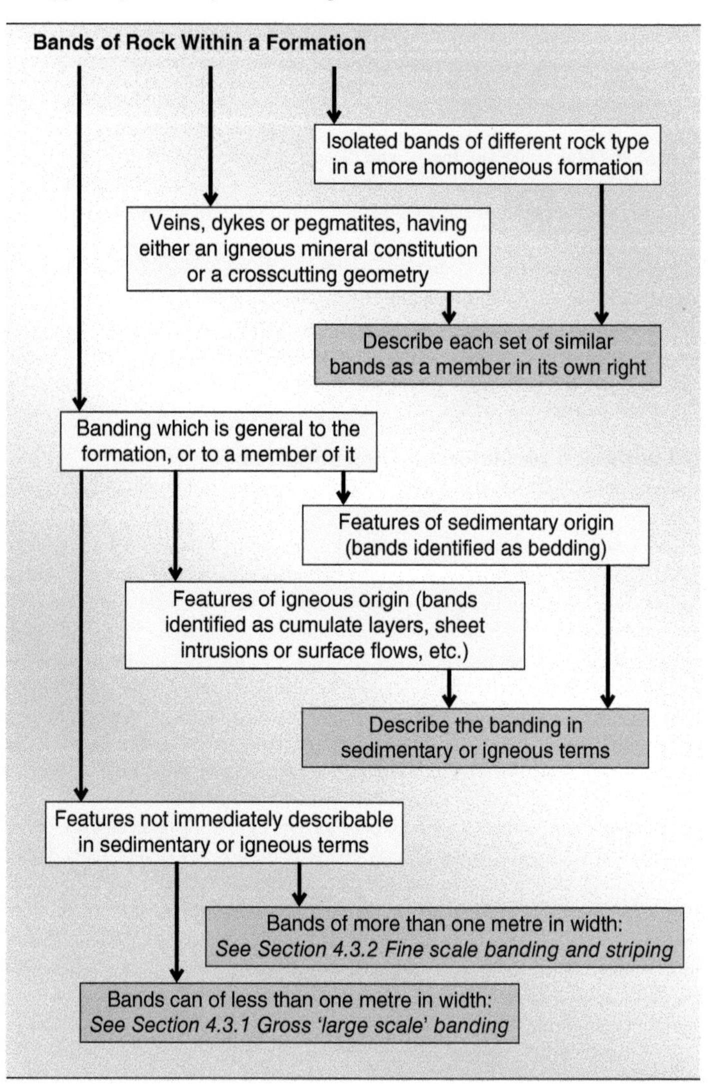

4.3.1 Gross 'large-scale' banding

Here we discuss banding at a scale of one metre or more that cannot be characterized completely in igneous or sedimentary terms. Most bands of this size do have a pre-metamorphic origin and provide information about the pre-metamorphic geology, though their current form may have been substantially modified by metamorphism and deformation. Obvious exceptions to this are veins, dykes, or sills intruded during metamorphism (e.g. Figure 4.3). Smaller-scale compositional bands, at sizes of one metre down to individual mineral grains (see Section 4.3.2), may be superimposed on the gross banding.

Although the sheet-like three-dimensional forms of rock 'bands' may have existed in an original sedimentary or igneous sequence, substantial alteration of their shape is normally only accomplished

Figure 4.3 *Later cross-cutting examples. (a) An early intrusion into a paragneiss is tightly folded (axial planes vertical on image), and is cross-cut by a later, undeformed dyke; (b) well-banded paragneisses are obliquely cut by a pegmatite which has itself been boudinaged. Both photos from Ticino, southern Switzerland. (photos Mark Caddick).*

when considerable deformation accompanies metamorphism. The grain-scale texture within any element of the banded rock can, however, be modified by metamorphism alone. If substantial deformation has occurred, there may be some bands that originated as cross-cutting intrusions but have had their angular discordances so reduced by high strain as to become indiscernible. Other bands may have begun as massive equidimensional xenoliths, faulted blocks, or isolated channels, and been strained into pancake-like forms. In whatever form, a large layer of contrasting composition still most likely represents an entity of intrusive or pre-metamorphic origin. In all cases, it is important to record (i) the existence of banding, (ii) any evidence of the original nature of either the whole sequence or of components within it, and (iii) evidence of metamorphism, deformation, or both (see Section 4.3.3).

When first faced with an outcrop, you might wish to go through the following simple questions:

- *Is banding discernible?* Is it obvious, with bands having grossly different colour or mineral content? Is a more subtle banding present elsewhere, consisting of slighter variations in mineral proportions? Are there bands with different resistance to weathering, or different weathered colour, which may not otherwise be distinguishable? Is there variability from place to place in accessory mineral content, or in grain size, or in degree of grain aggregation, which may on careful examination turn out to define parallel bands of more constant character?
- *What features constitute and characterise the banding?* Do bands differ (i) based on the minerals that are present, (ii) in mineral proportions, (iii) in grain sizes, (iv) in grain shapes, (v) in degree of grain aggregation (e.g. Figure 4.4), or (vi) in several of these?
- *Do discrete homogeneous bands exist*, or do the rocks constantly change all the way through the sequence? Are certain compositional bands characterised by sharp boundaries and others by gradational boundary zones?
- *As the sequence is traversed, is there a change* (i) in the dominance of certain band compositions, (ii) in the relative thicknesses of different band types, or (iii) in absolute band thicknesses?
- *Do some bands show boudinage or 'pinch-and-swell'* (Figure 4.5)? If so, is this feature common to all bands of the particular composition, or does it exist only where neighbouring bands are of a particular compositional type?

Answers to all of these questions are worth recording, and complex variations can be recorded graphically with a logged section (see also Chapter 2.3.2).

Figure 4.4 *Examples of types of banding. (a) Deformed, quartz-rich segregations in a gneiss, Ticino, Switzerland; (b) coarse mineral banding, Syros, Greece; (c) mm-scale compositional banding is reflected in different mineralogies of each band, Tauern Window, Austria; (d) tremolite-rich vein with pale reaction selvages at outer margins, Syros, Greece. (photos Mark Caddick).*

Figure 4.5 *Boudinage/pinch and swell. (a) Pinch and swell structures in a marble, Naxos, Greece; (b) boudinaged low-grade contact metamorphosed sediments, Italy; (c) boudinaged and partly rotated competent layer within banded gneiss, Norway (photos (a) and (b) Mark Caddick, photo (c) Dougal Jerram).*

4.3.2 Fine-scale banding and striping

Further scrutiny is required in cases where compositional layering, or discontinuous streaking or striping, occurs on scales between those of individual grains and of approximately a metre wide, because there is a higher possibility of these having a purely metamorphic–deformational origin. These finer scales of banding can differ (i) in their potential origins, (ii) in the methods best used for their observation and recording, and (iii) in their uses to both authors and readers of outcrop descriptions. We emphasise that the choice of one metre in this classification is arbitrary and may be changed as appropriate. At these finer scales, detailed observations of the compositions, abundances, and grain sizes of each specific mineral, as well recording fine-scale textural features and gradations, are very important. Mass flux across lithological boundaries can sometimes be inferred by sampling detailed transects perpendicular to the interface, with lab-based chemical and isotopic techniques becoming increasingly available for application to natural samples.

As discussed previously, large-scale bands, even if somewhat changed during metamorphism, were probably either distinct entities before metamorphism or were intruded during or after this metamorphism. At smaller scales, fluid-assisted chemical diffusion and dissolution–reprecipitation commonly operate so intensely during metamorphism that they can drive wholesale chemical changes between adjacent bands in an attempt to attain chemical equilibrium. As metamorphism progresses, rock and mineral compositions are continuously altered by diffusion, reducing chemical potential gradients and effectively acting to homogenise strong initial heterogeneities. However, chemical equilibrium at one set of metamorphic conditions represents disequilibrium at a different set, so later phases of metamorphism at different temperatures are likely to lead to further transport and reorganization of chemical components (thus permitting modification of mineral assemblage and compositions). Some compositional bands may be enlarged by diffusive influx, while others may be diminished. New bands are likely to form as reaction-zones and selvages at contacts, and these can be useful for inferring metamorphic evolution (e.g. Figure 4.6, see also Chapter 6 – Section 6.3). There are also several processes

Figure 4.6 *Cm-scale, garnet-rich reaction selvages (pink) formed at the interface between contrasting compositions in a rock that experienced granulite-facies conditions, Manitoba, Canada (photo Mark Caddick).*

of stress- and melt-induced chemical segregation, which act on the scale at which considerable diffusion is possible, and may alter the nature of banding. Sophisticated tools are now available to calculate likely mineral–fluid equilibria as a function of pressure, temperature, and rock composition, but these generally rely on careful petrographic and chemical study and are thus most appropriate for follow-up work on samples collected from the field area. The primary job of the field geologist is thus to make and record careful and detailed observations and collect the most appropriate samples.

4.3.3 Recording banding in the field

Whether by graphic log or otherwise (see Chapter 2), repetition, cyclicity, grading, and any generally asymmetric features should be looked for, both at the scale of the banding and at a scale which spans many bands. Some bands may be grain-size graded, while others may become finer (or coarser) towards both sides of the band. Some may be graded in their mineral content (e.g. Figure 4.3), or in their refraction of a cleavage or schistosity (see Section 4.5). Some may have been deformed into cuspate shapes on one side only. If any such asymmetry exists, the banding probably originated as a sequence of rock layers, and the asymmetry probably indirectly represents a 'way up' (i.e. the younging direction of a sedimentary or volcano-sedimentary sequence). Keep in mind, however, that if primary grain size and compositional banding (such as in a turbidite) is metamorphosed, it is possible for the initially finer, clay-rich layers to coarsen much more than the initially coarser, quartz-rich domains, essentially inverting the fining upwards sequence. This can be revealed by carefully assessing both grain size variation *and* the composition of those mineral grains.

Where asymmetry is pervasive on the grain scale and through folding and obvious deformation structures, it is more likely the result of intense deformation (see also examples in Chapter 5, Section 5.2.1, which discusses deformed mineral grains and fragments). In extreme cases, the compositional bands themselves may be the result of a metamorphic process involving large-scale chemical transport at high temperature in the mid to deep crust (e.g. in gneisses), sometimes involving partial melting (forming migmatites). This is discussed in more detail in Section 4.4.5.

Examples chosen for description of fine-scale banding should illustrate the general banding in the formation, the extremes of its variation, and also cases where its nature is distinctly different from the general case. These can only be chosen after you have established the general nature of the larger banding. Again, it is often best to spend a significant amount of time assessing the outcrop from afar, then in more detail, and then again from distance before making detailed notes or choosing localities to sketch.

It may be useful to have a list of some common questions to address when working on finely banded rocks. For example:

- Are there continuous thin compositional bands, discontinuous lensoid stripes and streaks, or both?
- Are these found throughout the sequence, restricted to within specific compositions of larger bands, or found everywhere except in larger bands of certain type?
- Do individual bands or lenses have sharply defined edges or more gradational edge-zones? Or are their edges almost indefinable, with the band representing instead a general gradational patchiness in composition (e.g. Figure 4.7)?
- Do bands thin laterally and then break up into small boudins or lenses (e.g. Figure 4.5)? Do they thin to a discontinuous string of individual mineral grains?
- Are some bands broken by cross-cutting veins, while others are not (Figure 4.3)?
- Is there any apparent relationship between band thickness and composition? How much is banding due to differences in proportions of the same set of minerals, and how much to changes from one set of minerals to another? What are the proportions of different band types? Do these proportions vary either locally, or across a formation?

Recognised patterns of large-scale banding, together with reasonable assumptions about pre-metamorphic or intrusive origins, often permit simple deductions to be made about the nature of

Figure 4.7 *Examples of sharp versus graded banding. (a) Gradational and sharp contacts, Syros, Greece; (b) modal mineral variations, eclogite, Norway (photo (a) Mark Caddick, photo (b) courtesy of Hans Jørgen).*

banded formations. These may be used in producing detailed outcrop descriptions that can be useful in interpreting more regional-scale trends. At the finer scale, complications normally require specialist study and a hypothetical interpretation before any further deductions can be made. This may require thin section study of carefully collected hand specimens. So, in a non-specialist description, the task is simply to make accurate and detailed accounts of well-chosen and accurately located examples the nature of any banding.

4.4 The Development of Fabric, Cleavage, Schistosity, and Lineations

One of the most common and striking features of metamorphic rocks, particularly those formed during regional metamorphism, is the development of secondary planar features and lineations that result from deformation. These features are produced by reorientation of grains, breaking of grains, dissolution and reprecipitation of grains, and crystallisation or recrystallisation of new metamorphic minerals during metamorphism and deformation. The stress field that rocks experience during metamorphism at elevated pressure and temperature is generally asymmetrical, with higher stress in one particular orientation (generally referred to as sigma 1). New mineral growth responds accordingly (e.g. Figure 4.8), with any elongate or planar minerals (such as micas) growing perpendicular to this primary stress – the way that if you throw a pack of cards in the air, each individual card will eventually orientate itself horizontally on the ground, perpendicular to the orientation of the dominant stress that it feels (gravity). If many minerals align in parallel because of this, the rock will inherit a planar fabric (a foliation) parallel to this – analogous to the 'fabric' formed by the aligned cards in the pack. In many cases, this fabric will be in the same orientation as primary compositional banding (e.g. bedding) within the rock, but this is not a requirement and high angle intersections between a mineral-alignment foliation and bedding can often be found. Here we will consider the development of mineral fabrics and look at some of the key observations that can be made.

4.4.1 Mineral and shape fabrics

Mineral fabrics are formed by the preferred orientation of grains or grain aggregates along specific linear or planar directions (see Figure 4.9). The preferred orientation may be stronger for some minerals than others, which may also relate to growth of different minerals at different stages of metamorphism. In theory, there is a distinction between 'grain shape fabrics' (grains

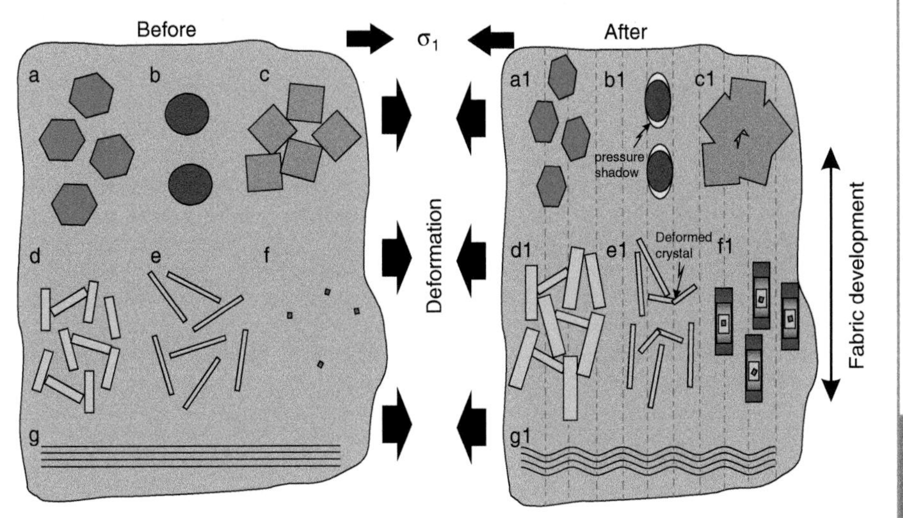

Figure 4.8 *Sigma 1 and fabric formation schematic. (a–a1) Grain/object deformation; (b–b1) grain/object deformation with new growth in shadow zone; (c–c1) pressure solution on grain boundaries; (d–d1) preferential growth of crystal aligned near parallel to strain; (e–e1) rotation and or deformation of crystals during deformation; (f–f1) preferential growth of crystals in alignment parallel to strain; (g–g1)folding of parallel layers (crenulation).*

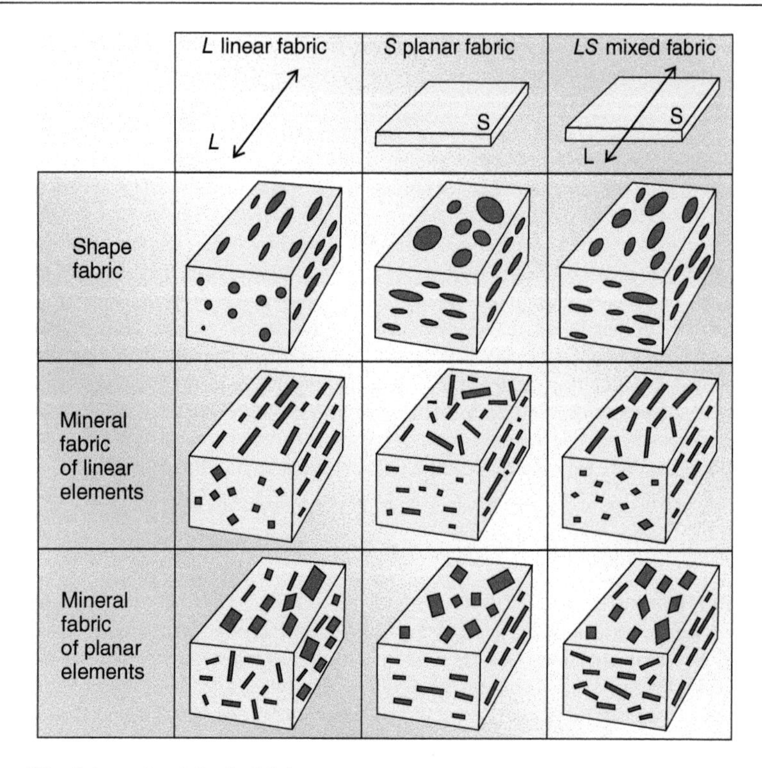

Figure 4.9 *Schematic of simple fabric symmetry types.*

with aligned external shapes regardless of the orientation of internal crystal axes) and 'grain orientation fabrics' (grains whose crystallographic axes are aligned, regardless of external shapes). In practice, such a distinction is too simplistic, the two factors are often inter-related, and orientation of mineral axes may be difficult to identify in the field. Many schists, for example, have a visible alignment of mica crystal directions and of both mica and quartz grain shapes. This can simply be called a 'mineral fabric' and can be recorded with a sketch, without worrying further about terminology at this point.

Patches of aggregated mineral grains can also have a preferred orientation (e.g. Figure 4.10). These can be described as having an 'aggregate shape fabric', or more simply 'shape fabric'.

4.4.2 Fabric symmetry

Mineral and shape fabrics, because they consist of discrete elements, can possess a so-called 'symmetry'. Any such fabric should be classified as 'L', 'S', or 'LS' type based on whether this symmetry is dominantly linear, planar, or mixed (see Figure 4.9). The same terminology may be used in a less rigorous way to describe whether other types of fabric are mainly linear, planar, or mixed in their resulting fissility (the way they break) or their visual effect. We will deal more with specific asymmetric patterns created in fabrics when considering shear zones in Section 7.3.2.

Figure 4.10 *Alpine rock with oriented mineral patches. Annotations highlight stretching and shortening directions. Note how different the textures look in different orientations, which is important to note when measuring deformation indicators (see also Chapter 7) (photo Dougal Jerram).*

4.4.3 Fabric origin and grade

Mineral and shape fabrics can have diverse origins. These can be primary, as in original alignment of minerals in a sedimentary protolith, or formed during subsequent deformation processes. You may find that different episodes of deformation have imparted different fabrics in a rock, and in this case additional textural information will need to be recorded in order to gain insight into the relative relationships of each fabric. When faced with a metamorphic fabric, evidence should be sought to address to two separate questions.

1. At what stage were the minerals or groups of minerals that constitute the fabric formed?
2. Were these elements originally formed with the fabric, or have they been rotated or otherwise deformed into their preferred orientation?

In meta-igneous rocks, alignment of grains or grain aggregates may result from original features such as magma flow, gravity settling, compaction, magmatic crystal growth, deformation associated with intrusion, or by later tectonic deformation. In metasediments, parallelism may be produced by sedimentation, diagenesis, and compaction, mimetic metamorphic mineral growth on a sedimentary fabric (i.e. crystal growth mimicking an original fabric), syn-metamorphic deformation, or deformation acting on pre-existing metamorphic minerals. Answers to questions regarding the origin of fabrics should always be sought from independent textural evidence at the outcrop scale, before considering relationships to tectonic structure (which is inherently more interpretive).

Where metamorphic minerals provide the fabric in the rock (e.g. the alignment of micas in a schist), they indicate the metamorphic grade of the rock at the time the fabric developed. This development may have been later than the time at which that metamorphic grade was first reached, as the processes of textural development and metamorphic mineral growth require time and can be decoupled. As the grade of the metamorphic rock increases, the extent to which new metamorphic mineral growth, diffusion, dissolution, and precipitation contribute to the observable fabric in the rocks changes, as evidenced by the different styles of fabric development in phyllites, slates, schists, and gneisses.

4.4.4 Slate fabrics, cleavages, schistosities, and lineations

Slates, phyllites, and schists have a directional fissility (an ability to split easily in a particular orientation) because of an alignment of mineral grains with good cleavage and/or a strongly sheet-like shape. This allows cracks to develop easily at the hand-sample scale by passing directly through mineral grains along their cleavages and between the individual crystal sheets. The phyllosilicates (such as micas, sericites, talc, chlorites, or serpentine minerals) are usually responsible for this, though minerals such as amphiboles (which also possess strong cleavage) can also contribute. To present almost continuous paths for the rock to split along, these minerals must either make up a substantial fraction of its composition, and/or be concentrated and aligned along laminations, crenulations, or pressure solution stripes (though in fine rocks this may not be visible).

The terms 'slate', 'phyllite', and 'schist' are generally used for rocks having a well-developed 'metamorphic' fabric (cleavage, Figure 4.11), though with different grain size and identity of minerals in each case. This in practice means that 'slate' may be used for all fine-grained rocks with any sign of fissility that is not due to an unambiguous sedimentary fabric. The terms 'cleavage' and 'slate' are used if the grains responsible for fissility are too small to be individually visible. In cases where such a rock splits along distinct and separate planes, it is said to have a *spaced cleavage* (see Figure 4.11b). If adjacent planes of potential splitting are so close that their character and spacing

Figure 4.11 *Examples of cleavage types. (a) Slate quarry, Penn Big bed, Slatedale, PA, USA. (b) Spaced cleavage in Windermere Group pencil slates, Banff National Park, Alberta, Canada. (c) Pencil slate in Nooksack Group, Glacier Creek Rd., Mt. Baker area, WA, USA. (d) Pressure solution cleavage, Ilfracombe Bay, Devon, UK (photos courtesy of Jim Talbot).*

are too small to be visible with a hand lens, this is termed a *penetrative* or *slaty cleavage* (Figure 4.11a). If a fine-grained rock has fissility in two or more directions (e.g. bedding and cleavage or with two or more cleavages), the rock may split easily into long thin pieces (like pencils) and be said to possess *pencil cleavage* (Figure 4.11c). If it breaks into flat pieces, it possesses *scaly cleavage*. In each of these cases, the orientations of the various cleavages should be recorded (if they are regular) together with any evidence of relative age (e.g. the crenulation of one cleavage by another). A fine-grained rock with only poorly developed fissility will not split into thin sheets, but will still fracture slightly more easily in one direction than another. This fracture with slightly preferred orientation is sometimes called *fracture cleavage*.

Rocks with grains that produce fissility and are large enough to be visible are schists, and their fissility is called schistosity (Figure 4.12). Different schistosity types, equivalent to the different cleavage types, are not normally named. However, a note (normally accompanied by a sketch) should be made of the type of fabric to which the schistosity corresponds, if possible. Many of the schists are formed by muscovite and biotite, but graphite and chlorite schists are also found, with variations between these generally reflecting the rock's composition and thus its protolith. In addition to the elongated or platy minerals within schists, characteristic isolated crystals such as garnet and andalusite can often be found, and in such cases the name of the schist can reflect this, e.g. garnet schist, staurolite schist, etc. (see Chapters 3 and 5 for more information).

Phyllites (e.g. Figure 4.13) have a grain size approximately on the borderline between 'slate' and 'schist'. Phyllites contain fine-grained mica flakes preferentially orientated along cleavage surfaces, and these crystals reflect light with a sheen called *phyllitic lustre*. However, this characteristic flakiness and sheen can only develop in rocks of particular, generally highly micaceous, composition. This is typically associated with broadly pelitic protoliths. 'Phyllite' is therefore better used as a rock-type name, and not, in the manner of 'slate' and 'schist', as a general grain-size category for fissile rocks regardless of composition.

Lineations are often found on cleavage planes (Figure 4.14). The intersection of bedding and a cleavage can impart a strong intersection lineation (Figure 4.14a), as can the intersection of two cleavages of different orientations (Figure 4.14b). Such structures are useful to identify and measure

Figure 4.12 *Example of schistosity as highlighted by crenulated shiny micas, Mull, Scotland. (photo Dougal Jerram).*

Figure 4.13 *Example of a phyllite, NW Scotland. (photo Dougal Jerram).*

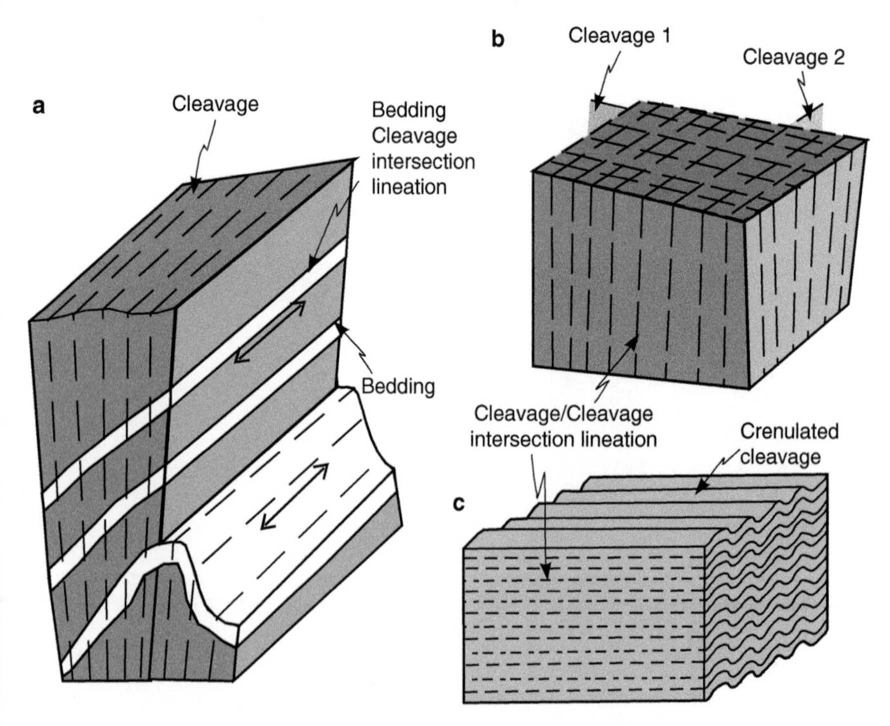

Figure 4.14 *Outline of lineation cleavage relationships. (a) Bedding/cleavage intersection lineations within cleaved folded beds; (b) cleavage/cleavage intersections; (c) cleavage/cleavage intersections in crenulated example.*

as they can provide important information about the structure and deformation history of your field area. A crenulation cleavage (Figure 4.14c and Figure 4.15) is formed due to the interaction of several fabrics that are themselves formed by two or more deformation events. In this case, an original planar cleavage, which formed as a response to deformation in one orientation, is subsequently kinked and deformed (crenulated) by deformation with a different orientation (see Figure 4.15). In some

Figure 4.15 *Examples of crenulation cleavage. (a) Well-preserved ripples in muscovite-andalusite schist. Hinges of millimetre-scale crenulation folds can be seen in the upper part of the image. Arkaroola, South Australia (image courtesy of Christoph Schrank). (b) Crenulated schist surface, Mull Scotland. (c) Large-scale crenulated bands, Namibia (photos (b) and (c), Dougal Jerram).*

instances, the mineral fabrics introduced in Section 4.4.1 help to contribute to a cleavage surface, and individual crystals or crystal aggregates can often be seen forming discrete rods or stripes, orientated along these cleavage planes. Sometimes, these rods can give an elongated 3D structure to a compositionally developed schistosity cleavage. In reality, many rocks experience multiple protracted phases of deformation, which can complicate the final cleavage configuration significantly.

In the field, such fabrics should be named, and at chosen localities they should be sketched to show whatever detail is visible, taking care to include a scale bar and to record orientations. Any early fabric, perhaps crenulated or cut by later pressure solution stripes, should be characterized in as much detail as the late fabric. The sense of displacement across crenulations, the sense of rotation from early to late fabrics, and their intersection directions should be under constant scrutiny. Changes of sense, or of intersection direction, should be located on maps and sections (see also Chapter 2 Section 2.3.3).

4.4.5 Gneiss banding and migmatites

At high grades of metamorphism, coarse banding can develop which is often synonymous with folding and ductile deformation. As conditions become much hotter and with increased pressure, alternating darker and lighter coloured bands, called *gneissic banding*, develop (Figure 4.16). A rock exhibiting this banding is called a *gneiss*. This can be a further development of pre-existing compositional layers, such as bedding, which were present in the protolith (i.e. lighter quartz-rich and darker organic-rich shale beds). At high grade, these layers recrystallise into quartzites and mica-rich bands. In other cases, wholesale differentiation of material has taken place during metamorphism, leading to segregation into felsic light bands and mafic dark bands. In extreme cases, the rock starts to partially melt and the melt segregates efficiently into bands. Some of the possible reactions that would lead to this melting were shown in Figure 3.17. Where the rock has clearly partially melted, it is known as a migmatite (Figure 4.17) and is likely to show distinctive textures comprised of well-defined, alternating layers of:

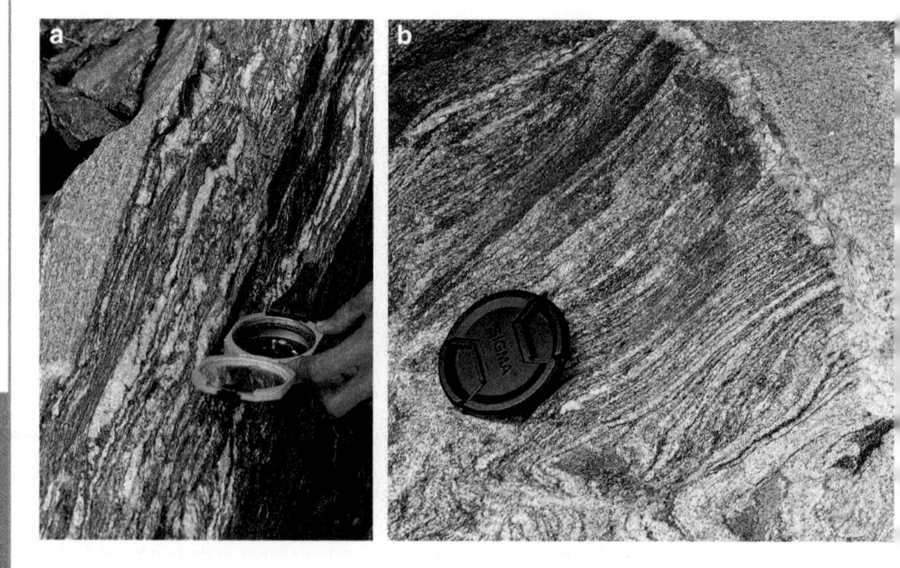

Figure 4.16 *Gneisses and migmatites: (a) mm to cm-scale irregular and folded leucosomes and melanosomes, migmatite from Montana, USA; (b) planar migmatitic segregations in a gneiss cross-cut by a granitic dyke in top right of image, Naxos, Greece (photos Mark Caddick).*

- *Leucosome*. Broadly granitic compositions dominated by quartz and feldspar.
- *Melanosome*. Dark bands of mafic-rich material such as amphiboles and biotites.
- *Mesosomes*. Rock compositions between the melanosome and leucosome (see Figure 4.17). The mesosome may represent the unsegregated composition, in which case it is termed the paleosome.

There is no sharp distinction between some banded gneiss and migmatites, and terms such as migmatitic gneiss may be used.

Figure 4.17 *(a) Large, planar segregations in a migmatite, Montana, USA. (b) Complexly deformed migmatite, Limpopo, South Africa (photos Mark Caddick).*

4.5 Refraction, Kinking, and Shearing of Fabrics

The systematic variation in orientation of a fabric as it traverses a band of rock is known as *refraction*. The band in question may be a *compositional* entity (a bed, for example) or it may be a *deformation band* or *zone* which is not compositionally defined.

4.5.1 Refraction

The strains experienced by bands of different composition when a rock deforms will differ, according to the competence of the bands. This gives rise to variation in the orientation at which a fabric develops. Planar fabrics cut the most competent bands nearest to their perpendicular. If the fabric is responsible for a cleavage, this produces *cleavage refraction* (Figures 4.18 and 4.19). This can be useful in showing *grading* and the *way up* of deformed, fine-grained, graded beds. The term *refraction* is often restricted to this usage.

4.5.2 Shear and kink-bands

Shear type deformation zones are not essentially demarcated by specific compositions, nor are they geometrically limited by a pre-existing fabric. Several fabric relations may exist in shear zones (e.g. Figure 4.18; see also Chapter 7). A pre-existing shape fabric will be reoriented according to the finite strain. A pre-existing lamination will be reoriented as a material entity. A new mineral,

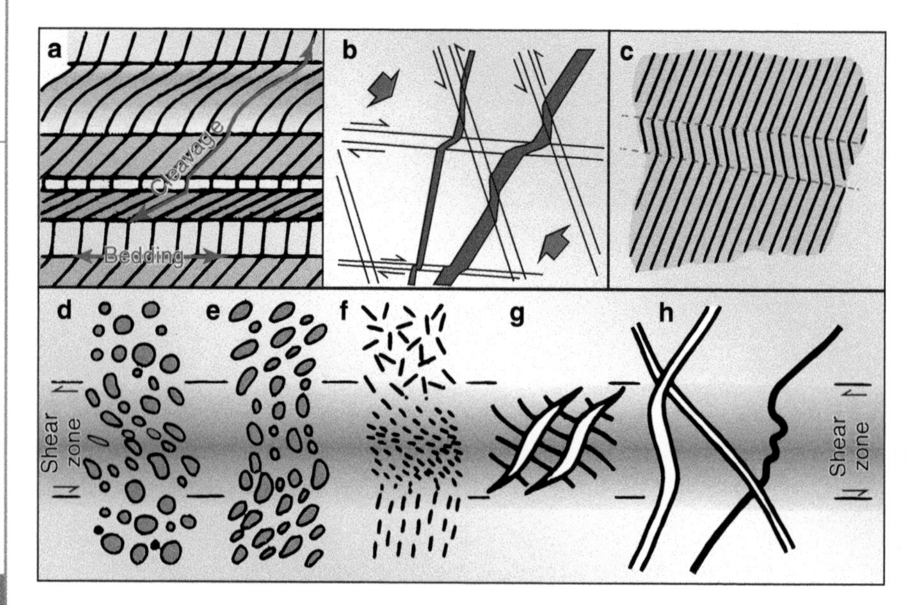

Figure 4.18 *Refraction, kinking, and shearing. Top: (a) Refraction, showing the traces of a fabric (e.g. cleavage) through horizontal bands (e.g. bedding), one of which is graded. (b) A conjugate set of shear zones (demarcated by black lines) off-setting dark brown compositional bands. The black arrows indicate displacement sense across individual shear zones, and the common principal direction of shortening. (c) Traces of fabric (black lines) through a kink-band (demarcated by red lines). Bottom: A sinistral shear-zone passing through unaffected rock, showing (left to right): (d) A new-shape fabric. (e) A reworked shape fabric. (f) A new mineral fabric. (g) A spaced cleavage (e.g. pressure solution stripes) cut by sigmoidal extensional veins. (h) Passive folding of pale veins and active folding of a dark vein.*

Figure 4.19 *Example of cleavage refraction in metasediments, Namibia (photo Dougal Jerram).*

pressure solution, or slaty fabric may develop. The strain may not die out completely outside the sheared zone, but simply reduce to a lower value. Sheared fabrics and shear zones exist at all sizes from the sub-microscopic to many kilometres wide, and as such the larger-scale shear zones can make for very important metamorphic zones/boundaries. Shear zones of this nature are dealt with in more detail in Chapter 7.

A *kink-band* has some similarity to a shear zone, except that a previous fabric plays an active role in its development. This is shown by the sharp angular change in fabric orientation at the edge of the band, and the roughly equal angle between this edge and the fabric inside and outside the band (Figures 4.18 and 4.20).

4.6 Deformation Fabrics and Folds

There are some important ways that deformation fabrics develop in relation to folding, and also how they are affected by folding. The formation of fabrics can occur during a folding episode, after there has been significant folding, or before a subsequent folding and deformation event (e.g. Figure 4.21). It is important to gain an understanding of the different relationships that exist between folds and fabrics when faced with both in the field.

Consistent fold patterns provide a basis for structural correlation. Any clear relationship among metamorphic minerals, their resulting fabrics, and folds, can therefore be extremely important. Folds and fabrics should be measured at the same locality using the normal structural features, which can be recorded (e.g. Chapter 2 Section 2.3.3). Folds can be conveniently classified as *active* or *passive*, the former requiring rheological contrasts within the folded material that exert a mechanical influence on the folding, the latter lacking these contrasts. *Passive folds* can occur locally in shear zones, and more generally where masses of heterogeneously deformed rocks lack a consistent orientation of anisotropy in their physical properties. Development of a fabric during *active* folding would be supported by evidence of an axial-planar cleavage or neutral point relationship. A neutral point forms because, although the majority of the rock is experiencing shortening, a small zone on the outside of fold hinges of more competent layers will experience very localised stretching (similar

Figure 4.20 *Kinked foliation, Boscastle, Cornwall (photo width ~1.5 m) (photo courtesy of Jim Talbot).*

to strain shadows around porphyroblasts; see Chapter 5) – the neutral point describes the locality at which stretching passes into compression (see Figure 4.22).

A difference in the time of development of fold and penetrative fabric can be established where a fabric is folded, or where a fabric cuts across and disregards (save for refraction) a previous fold (Figure 4.22). However, it is quite possible that no consistent pattern of folds or fabrics exists, or that folds are purely passive features. In general, the higher the metamorphic grade, and the more passive the rock units, the less likely there is to be simple and consistent folding. *In the field*, it is necessary to *show whether a consistent relationship exists or not* between fabric and structures.

4.6.1 Fabrics forming with folding

The relationship known as 'axial-planar' (Figures 4.21 and 4.22a) is a common consequence of planar fabrics and folds being produced during the same deformation event. In reality, the fabrics that form during folding are rarely perfectly axial-planar, and fabrics can fan through bands of different competence (e.g. Section 4.5.1 and Figures 4.21 and 4.22). The name 'axial-planar' may therefore be strictly inaccurate, though the term is in common usage and is conceptually important, as a planar fabric that develops synchronously with folding will lie close in orientation to the fold's axial plane. Therefore, the fabric cuts banding on fold limbs with the same sense of asymmetry as it would if truly axial-planar. *The practical importance of this is that, where it can be demonstrated to exist:*

1. *It justifies correlating the episodes of formation of fabric and of folding.*
2. *Banding/fabric sense can be used to extrapolate from smaller to larger structures, in the same way as parasitic folds.*

4.6.2 Superimposing folds and fabrics

A planar fabric that developed later than a fold will be superimposed on the fold geometry, subject to normal refraction and fanning around areas of contrasting ductility. This might conceivably produce a relationship that is again axial-planar (to the earlier fold) if there is near coincidence of strain

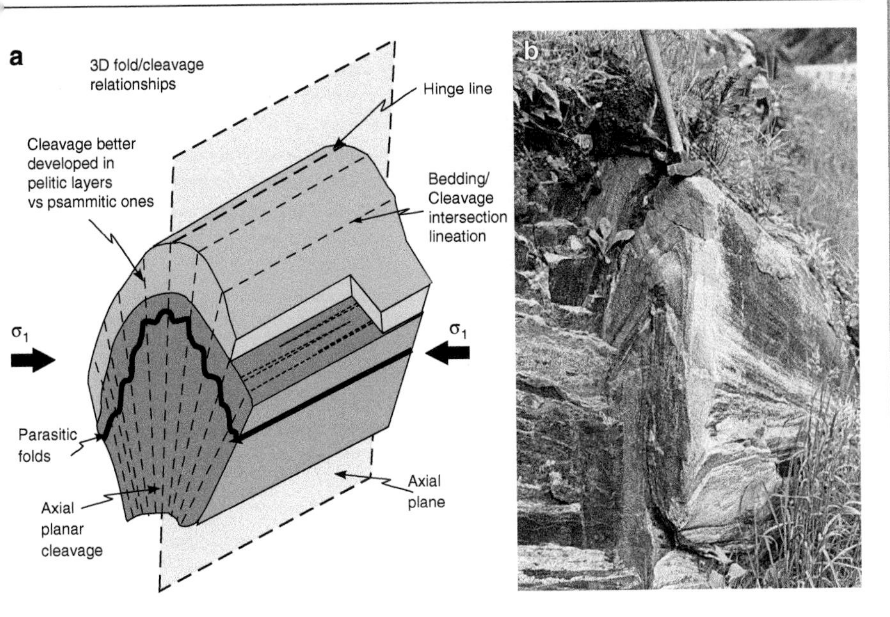

Figure 4.21 *(a) Fold and cleavage relationships in three dimensions. (b) Example of fold with 3D outcrop (photo courtesy of Jim Talbot).*

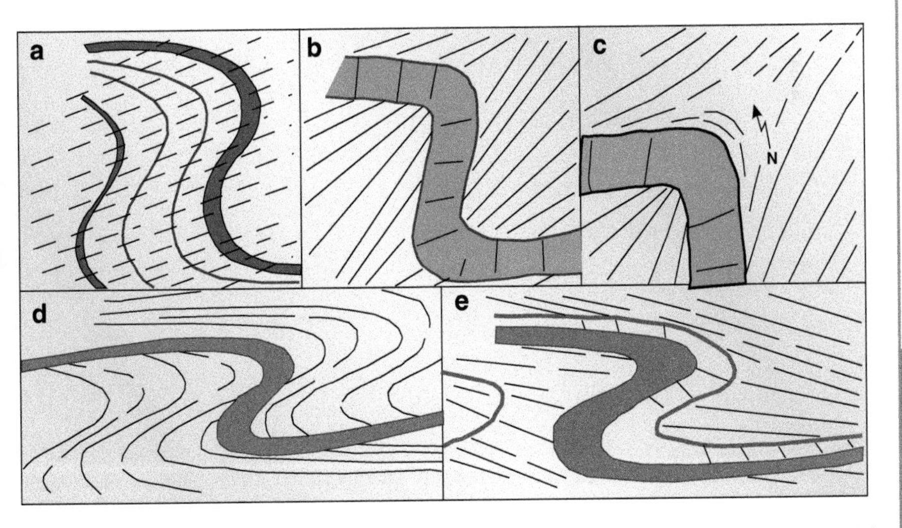

Figure 4.22 *Fabric/fold relationships on profile sections. Thin lines represent fabric traces. (a) Axial-planar. (b) Axial-planar with fanning. (c) Neutral point (N). (d) Fabric earlier than folding. (e) Fabric later than folding showing refraction (one possible configuration).*

orientations of the two deformation episodes. If the potential new fabric orientation lies only a few degrees from an existing one, it is normal for the earlier fabric to be redeveloped instead of a new fabric developing. Thus, two deformation episodes can result in only one apparent fabric, axial-planar to one generation of folding. In this case, the episodes cannot be discerned on the basis of evidence from either the planar fabric or the fold. Ideally, you would be able to see and measure the difference between two or more folding events, where the shortening directions that formed them significantly differed. Where this is less clear, you may see other evidence of multiple deformation events by recognising textural relationships including crenulation cleavage, refolded folds, and the rotation of inclusion trails in porphyroblasts (see Chapter 5 Section 5.2.1).

4.6.3 Only correlate like with like

When using relationships between fabrics and folds, it is important only to correlate like with like. This means, for example, that mineral fabrics belonging to a particular deformation episode should not be correlated with shape fabrics which record the finite strain accumulated through all episodes. Fabrics should not be correlated on orientation if they are produced by minerals of quite different metamorphic grade.

In terms of folds, it is important to realize that fold orientations will only be constant over those rock masses experiencing the same pre-folding rheology contrasts and stress orientations. In particular, some folds may develop by buckling veins or intrusions, others by buckling banding, and some by buckling an earlier fabric. These may all share an approximately common axial-planar orientation and axial-planar fabric, but they may differ in fold asymmetry, sense to banding, and hinge direction. They cannot, therefore, all be used together to establish larger structures. Veins, in particular, are notorious for producing folds oriented differently from those of banding or fabric.

Those features which appear to actively create the folding, and those which appear to be passively folded by it, should be noted.

UNDERSTANDING TEXTURES AND FABRICS 2: METAMORPHIC CRYSTALS, PSEUDOMORPHS, AND SCATTERED ENTITIES

Boulder of a garnet amphibolite with large garnet porphyroblasts, Gore Mt, New York, USA. Mark Caddick for scale.

5

UNDERSTANDING TEXTURES AND FABRICS 2: METAMORPHIC CRYSTALS, PSEUDOMORPHS, AND SCATTERED ENTITIES

This chapter deals with easily identifiable entities such as metamorphic crystals, deformed crystals, and larger bodies including veins, all of which are commonly found within metamorphic rocks. The first section discusses some of the common metamorphic textures associated with the growth of visible metamorphic minerals. These can be found as isolated grains or groups of crystals and can sometimes exhibit useful growth textures and deformation textures that are worth noting. Pseudomorphs – patches having the shape of individual grains of one mineral while consisting of one or several other minerals – are considered, with their implications for deciphering overprinted metamorphic events. The next section then deals with lens or pod-like objects in rocks that have been deformed.

5.1 Recording Metamorphic Textures

Once the general structure of an outcrop has been examined, with any banding and fabrics noted and measured, and their relationships sketched, it is time to turn to your attention to texture. In some cases, this process will be completed along with fabric analysis in finely banded or laminated rocks, where the main rock texture is brought out solely by its fabric. In many other cases, a detailed investigation of the textural elements of key parts of the outcrop will be needed to help determine metamorphic crystal type, growth and deformation histories, and additional finer details.

All textural matters are fundamentally geometrical, and are recorded best in sketches. Many of the textural terms used in metamorphic rocks, including 'porphyroblast', 'porphyroclast', 'radial', 'tabular', and 'acicular', are useful rock descriptors, but have much greater value if accompanied by appropriate diagrams. A properly annotated sketch will show relationships with the minerals adjacent to those for which such a term might apply, revealing relative grain-sizes, habits of other minerals, degrees of mineral association, patchiness, directionality, etc.

In the field, as a rule:

1. Always sketch.
2. Always annotate with names of minerals.
3. Always include a scale bar.
4. Always record the attitude of the sketched surface, and the pitch of features on it.

We have given some examples of field notebook sketches in Chapter 2, and emphasise here that different styles and sizes of sketch are appropriate for recording different types of information. You will develop your own style of recording information diagrammatically, but all useful diagrams take quite some time to complete – so be prepared to spend several minutes on each. Alongside or within larger sketches you can include many individual sketches of textural features, as side notes in logs, expanded boxes, or as additional annotations as part of a larger, composite outcrop sketch. Some examples of simple texture sketches are given in Figures 5.1 and 5.2. The structure, fabric, and texture of metamorphic rocks can be complex on a number of scales, and continuing with a theme

The Field Description of Metamorphic Rocks, Second Edition. Dougal Jerram and Mark Caddick.

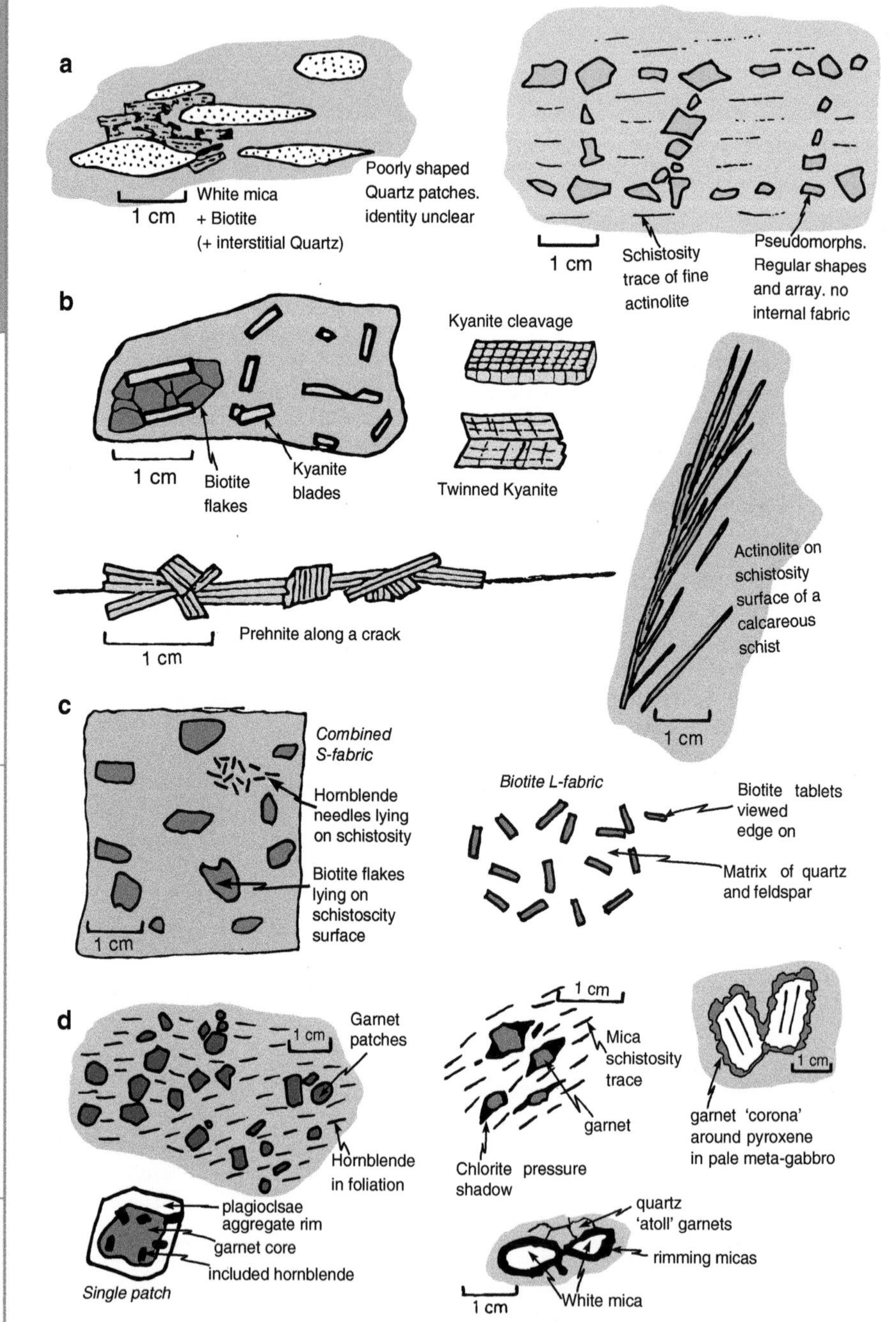

Figure 5.1 *Field sketches of textures from which the sequence of mineral growth and deformation can be inferred. (a) Shape, size, and distribution of compositional and mineralogical patches. (b) Minerals growing with specific habits: kyanite, prehnite, actinolite. (c) Some mineral fabrics. (d) Examples of mineral associations and inferred time relationships, all involving garnet.*

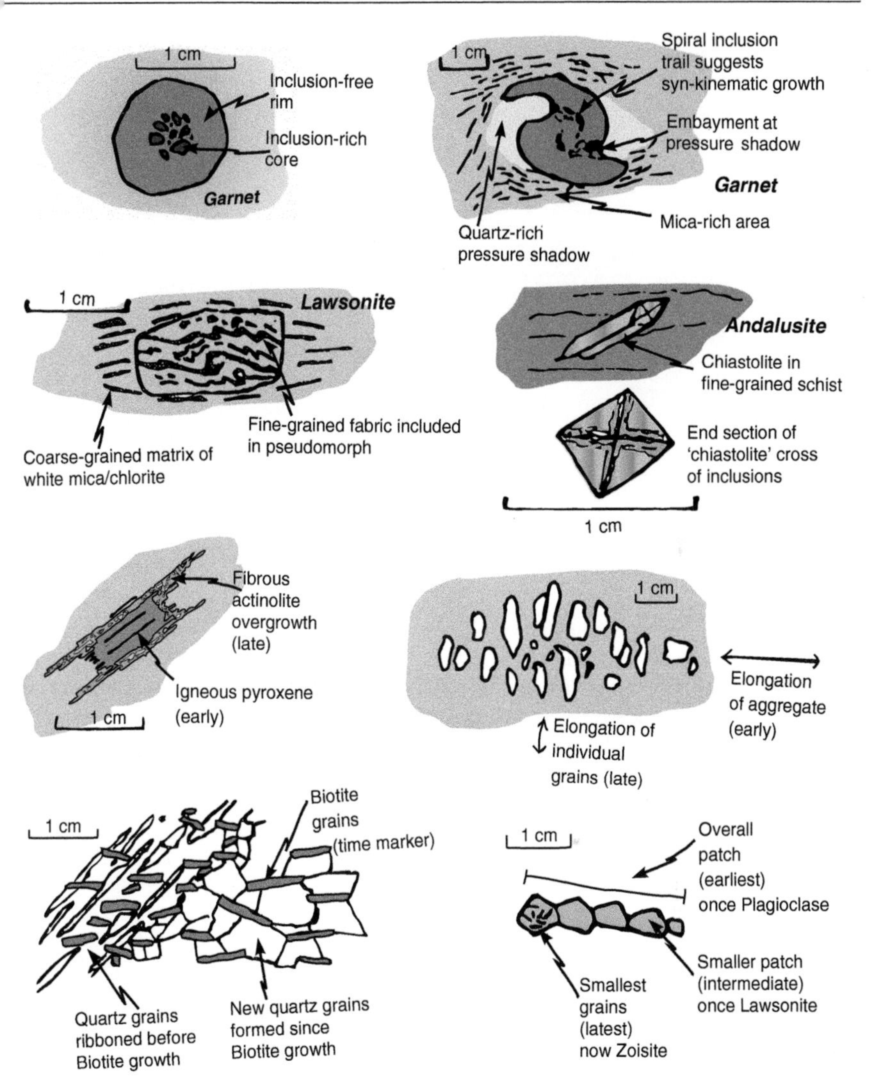

Figure 5.2 *Sketches of textures that show relationships between metamorphic mineral growth and deformation, revealing relative timing of events.*

of top down (beginning with broader scale features before zooming into detail), we shall first consider some of the larger textural variations that are often found in metamorphic rocks before focusing on grain/crystal scale observations.

5.2 Metamorphic Crystal Growth and Porphyroblasts

As a metamorphic texture develops and matures, new minerals grow at the expense of some of the existing minerals, forming new mineral assemblages and crystal sizes and shapes. When fully equilibrated, these assemblages reflect the pressure and temperature conditions that the rock mass resided

Figure 5.3 *Garnet porphyroblasts (orange) and pseudomorphs after lawsonite (cream) in a white-mica and amphibole rich blueschist, Syros, Greece (photo Mark Caddick).*

at for sufficient time for the equilibration to occur. The protolith composition, pressure and temperature conditions, efficiency of diffusion of nutrients through the rock, presence of metamorphic fluids, and additional factors, may allow some metamorphic minerals to grow into large crystals, relative to their finer grained ground mass (e.g. Figure 5.3).

In igneous rocks, scattered mineral grains that are distinctly larger than those of an intervening matrix are called *phenocrysts*. In metamorphic rocks, such scattered crystals that have grown much larger than those in the matrix are called *porphyroblasts*. These porphyroblasts can be reminiscent of larger phenocrysts in igneous rock textures, and care must be taken to identify the large crystal present as a true metamorphic crystal when looking at meta-igneous rocks (see Section 5.2.2). The mineralogies of porphyroblasts are often more readily identified than matrix minerals and they can thus provide useful information about metamorphic grade, as well as insights into the composition of the protolith. Some examples of porphyroblasts are given in Figure 5.4. It should be noted that porphyroblasts may grow in orientations along with the strain in the rock (common in growth stages that occur during early to peak strain), or in apparent random distributions (common in contact rocks and late-stage mineral growth).

5.2.1 Deformed mineral grains and fragments; porphyroclasts, augen, flaser, and large mineral grains

Relict (inherited) grains/crystals that are larger than others in a deformed rock are known as *porphyroclasts*. The key difference between a porphyroblast and a porphyroclast is that porphyroblasts are larger than the surrounding grains due to differences in metamorphic crystal growth, whereas porphyroclasts are bigger than the surrounding texture due predominantly to grain-size reduction of the host during deformation. It is generally possible to demonstrate that some deformation/shear has occurred around porphyroclasts, or that they have been rotated relative to the matrix. Lensoid, spindle-shaped, or discus-shaped compositional patches with pointed ends are known as either *flaser* (streaks) or *augen* (meaning 'eyes', plural). Augen are produced by deformation, with the sharp ends around a wider centre resulting from strain being localized in their rims (Figure 5.5). If such a distinct contrast between less strained core and

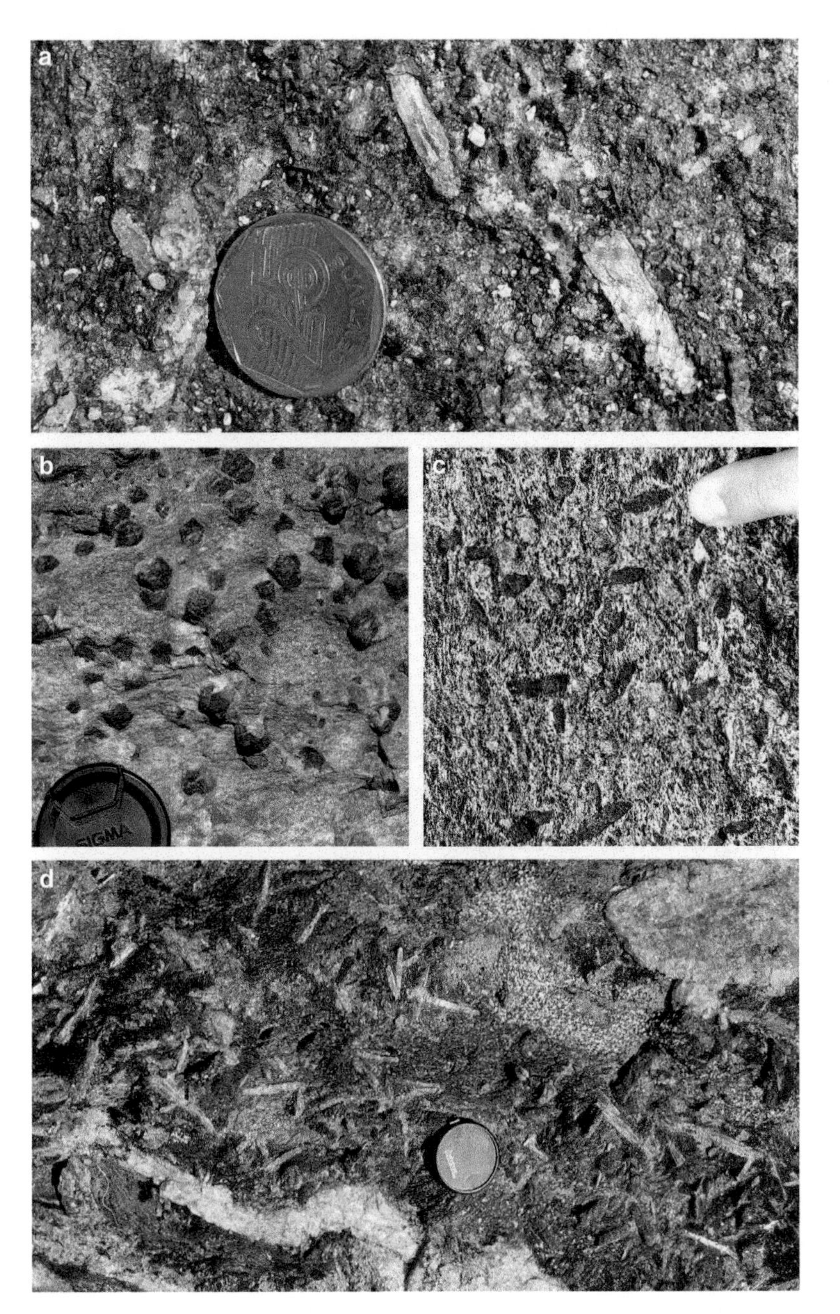

Figure 5.4 *Examples of porphyroblasts. (a) Kyanite-garnet paragneiss from the Ediacaran Buzios succession within the Cabo Frio Tectonic Domain, Buzios, RJ, SE Brazil. (b) Garnet porphyroblasts in a fine-grained, mica-rich schist, Syros, Greece. (c) Amphibole and garnet porphyroblasts in metasediments, Ticino, Switzerland. (d) Kyanite, garnet, and staurolite in amphibolite schist from the Matchless Amphibolite of the Damara Orogen, Gorob Mine, Namibia (photos: (a) courtesy of Isabela Carmo; (b) and (c) courtesy of Mark Caddick; (d) courtesy of Susanne Schmid).*

Figure 5.5 *Example of an augen gneiss, Brazil (photo courtesy of Isabela Carmo).*

Figure 5.6 *Flaser fabric in a quartz, feldspar, and biotite-rich gneiss from Virginia, USA (photo by Dougal Jerram – sample courtesy of R.J. Tracy).*

more strained rim is lacking, a thinner flaser shape is produced (Figure 5.6). The flaser fabric contains small streaked-out porphyroclasts in a finer sheared matrix, which often forms the typical background texture of mylonitic/sheared rocks (the term flaser is more commonly used in sediments to describe sandy lenses in a muddy matrix). Augen (or flaser) may have started as unusually large grains (clasts, phenocrysts, or porphyroblasts), or as grains that were the same size as other minerals until deformation and metamorphism occurred and preferentially reduced grain sizes.

Augen porphyroclasts can produce quite distinctive textures. These are patches of distinct composition, each of which was at some time (in many cases) a single grain and now has the appearance of an 'eye' shape on a rock surface. The shape consists of a more equidimensional core, mantled by (or having tails of) a deformed aggregate of mineral grains which, although streaked out, are still part of the compositionally defined patch. The core may be a single crystal, a remnant of the grain that originally constituted the patch. The mantling grains may be either the same mineral, or a new mineral with similar chemical composition. In three dimensions, the entire patch may be spindle-shaped, discus-shaped, or something between the two. Because the cores may preserve an early structural state and orientation, and are mantled by material developed during more recent deformation, augen can sometimes be used in correlations of simple structural and metamorphic rock histories. Asymmetry of the mantling minerals can be used to infer the direction of shear (e.g. Figures 5.7 and 5.8), as can any observation of partially rotated crystals in which a previous fabric is preserved

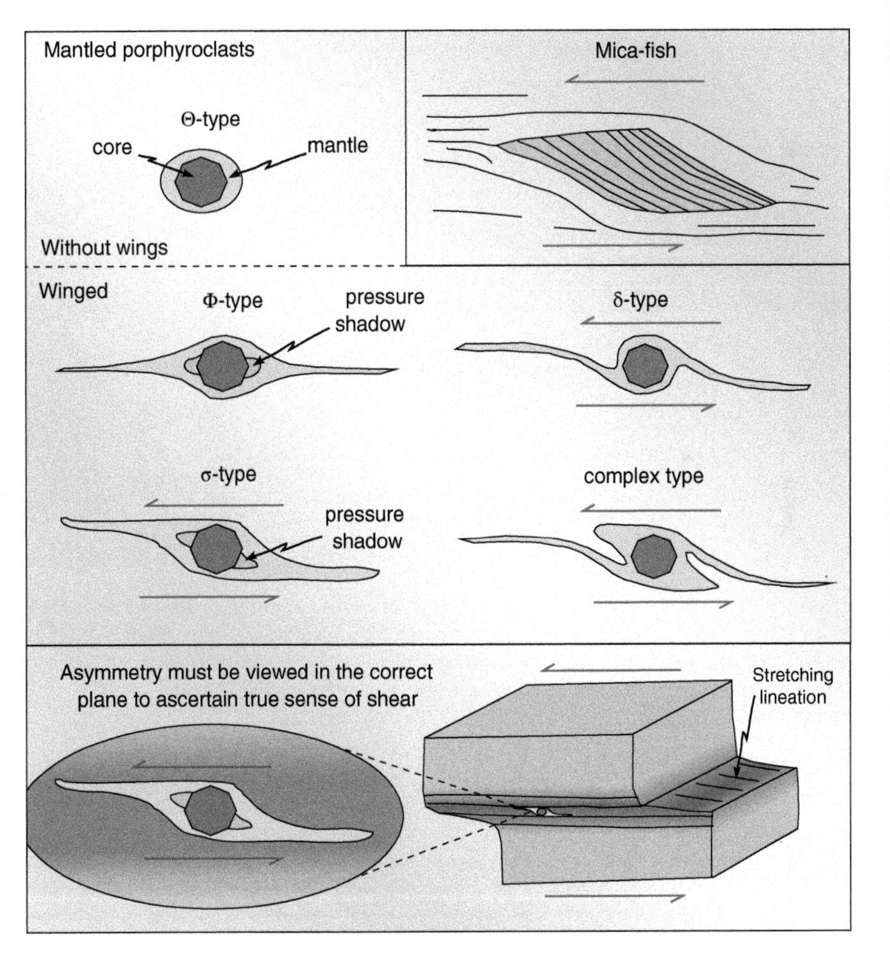

Figure 5.7 *Sketches of early metamorphic (or inherited) crystals that can reveal information about deformation through the geometry of a mantle around them or through their direct rotation and deformation (e.g. mica fish).*

Figure 5.8 *Examples of rotated augen (a) and (b) sheared feldspar porphyroclasts in gneiss. The delta clast above the marker coin indicates dextral shear sense. Western Grenville Province, Ontario, Canada. (c) Asymmetric hornblendite ball in Lewisian gneiss. Achmelvic, NW Highlands (photos: (a) and (b) courtesy of Christoph Schrank; (c) courtesy of Jim Talbot).*

as mineral inclusions within large grains but is not parallel to the primary fabric of the rock (e.g Figure 5.9, Figure 5.10). In some instances, it can be demonstrated that crystals grew either before during, or after deformation events that form the main fabric of the rock (e.g. Figure 5.9). A classic example that can sometimes be seen within garnets is evidence of them being rotated during growth, which can be identified by sinusoidal inclusion trails within porphyroblasts (e.g. Figures 5.9b1 and 5.10). This is more commonly seen at the thin section scale than in the field, but it is always worth looking out for!

During deformation, the presence of larger, more competent grains can induce localised pressure variations within a rock, with low-pressure zones (so-called pressure shadows) directly adjacent to some crystals (see Figure 5.7). Different minerals can grow under these

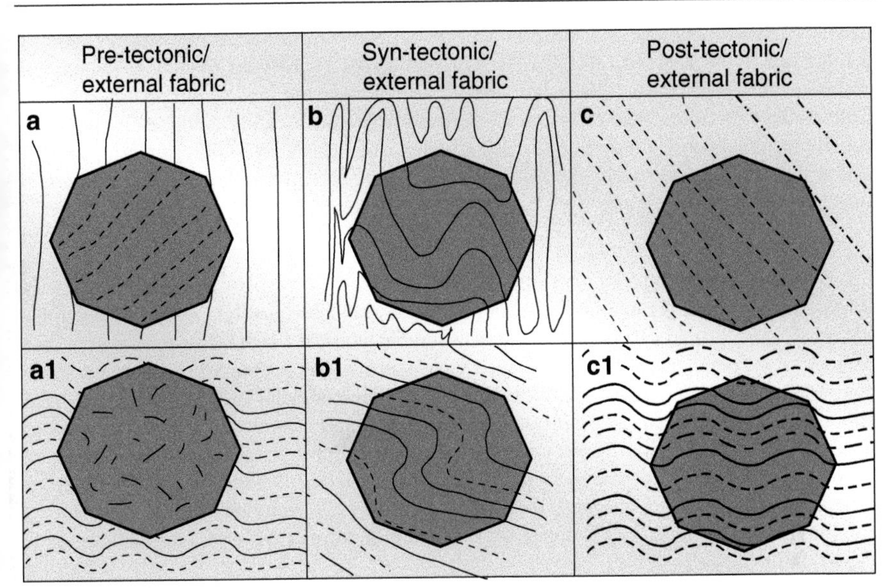

Figure 5.9 *Schematic of different stages of fabric development in relation to porphyroblast growth (for example garnet). (a) and (a1) porphyroblast formed prior to current foliation/schistosity (in (a), the porphyroblast contains a previous fabric; in (a1), it contains no initial fabric of note). (b) and (b1) porphyroblast grows during deformation event (in (b), an early fabric is preserved with less deformation than final fabric in host; in (b1) the porphyroblast grows and rotates during fabric development). (c) and (c1) porphyroblasts grow after the external fabric is developed.*

Figure 5.10 *Rotated inclusions in garnet compared to matrix. Example shows quartz and phengite inclusions in garnet, Syros, Greece (photo Mark Caddick).*

locally different conditions, or material can be dissolved from higher pressure regions and precipitated in the shadows (also discussed in Section 4.2.1). Pressure shadow differentiation around grains gives them 'elbows' or 'tails', producing a similar overall shape to augen. Unlike augen, pressure shadow materials are usually compositionally unlike their competent equidimensional cores. Fibrous or bearded growths may occur both within a pressure shadow and at the ends of augen, though exactly parallel fringes are generally restricted to pressure shadows (Figure 5.7).

5.2.2 Pseudomorphs

When porphyroblasts grew at one metamorphic grade and subsequently reacted to form a new mineral or suite of minerals at a different metamorphic grade, but kept the original crystal morphology of the porphyroblast, they are called *pseudomorphs* (e.g. Figure 5.11). These are important because they demonstrate a relationship between two mineralogically definable geological episodes, one in which an early mineral was created, the other in which it was replaced by later minerals. They are therefore important in the establishment of a geological history, a pressure–temperature path (both prograde and retrograde), and the potential importance of key burial stages (or even intrusive episodes in contact metamorphic examples). Identifying the minerals of the two episodes from their similarities of composition can be difficult, and this aspect should therefore be tackled last.

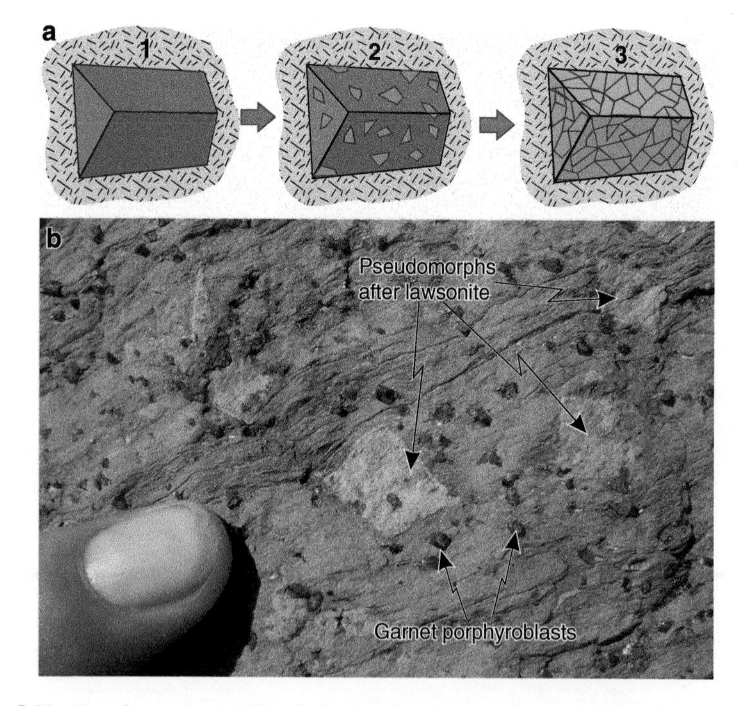

Figure 5.11 *Pseudomorphs. (a) Sketch showing development from original mineral in brown to new phase in green. (b) Example photo of pseudomorphs after lawsonite (note that some contain inclusions of garnet), Syros, Greece (see also figure 3.8) (photo Dougal Jerram).*

5.2.2.1 *Early mineral and the textural evidence*

Textural evidence can often be used to establish the stage in the rock's history at which the pseudomorphed mineral existed. In metamorphosed plutonic rocks, the earliest non-metamorphic texture is igneous. Is the pseudomorph an integral part of such a texture? If so, the mineral was igneous. In these cases, the chemical constituents in a phenocryst or coarse igneous crystalline texture have been subsequently used to create a metamorphic mineral or mineral assemblage with the same morphology as the original igneous rock (e.g. Figure 5.12).

In coarse metamorphic rocks, the latest texture is that of the current mineral assemblage. That is to say that mineral assemblages reflecting earlier stages of metamorphism have been successively replaced by later assemblages. The latest assemblage often forms during retrogression, though elements of higher-temperature mineralogy might persist. In such cases the pseudomorphing is generally relatively late, and is probably a retrograde replacement. Such situations can be useful to help to work out the particular $P–T$ path that a rock may have followed during its prograde and retrograde phases. The example in Figure 5.11 highlights this: early-formed lawsonite crystals reflect high pressure, low temperature within a subduction zone, but are now pseudomorphed by a mineral assemblage that formed later, at much shallower conditions. The new minerals are far too fine-grained to be seen with the naked eye, but in aggregate they retain the shape of the original lawsonite crystal (see also Figure 3.8 for thin-section-scale images of these features).

These are the easy cases, where quite strict pressure–temperature limits are placed on possible pseudomorphed minerals by knowledge of the rest of their assemblage. In other cases, the early stage of the metamorphic history may have to be considered in terms of possible minerals, veining, deformation, and grade.

Phenocryst or porphyroblast? A common problem is the identification of large pseudomorphs in an otherwise homogeneous, finer-grained meta-igneous rock. If location, size distribution, orientation, or deformation state positively prove them to be replacement after porphyroblasts, then the problem does not arise. Examples of this are:

1. Pseudomorphs enlarged or concentrated along small fault-planes or joints.
2. Undeformed pseudomorphs defining a deformation fabric.
3. Pseudomorphs overgrowing a previous fabric.
4. Pseudomorphs occurring in hydrothermal veins as well as host rock.

Figure 5.12 *Gabbroic texture overgrown by green metamorphic pyroxene, Syros, Greece (photo Mark Caddick).*

These all show that a post-igneous process preceded or accompanied the early mineral, which is therefore metamorphic. If the location, size distribution, and orientation are explicable in igneous terms, their origin may be difficult to determine. Igneous minerals are usually better developed in coarse-grained igneous rocks such as gabbros, granites, and pegmatites. These situations are all possible sites for preferential growth of metamorphic minerals. In cases such as these, it is normally necessary to rely on pseudomorph shape, plus knowledge of igneous mineral compositions (e.g. original protolith relationships), to identify the early mineral.

5.2.2.2 Pseudomorph identification and questions about metamorphic history

In cases where the pseudomorphed mineral is part of the rock's early metamorphic history (i.e. it is not a phenocryst inherited from an igneous protolith), it may be possible to make definitive statements about its timing in terms of textures and fabrics without knowing exactly what the mineral originally was. However, with good crystal form of the parent mineral, and the new pseudomorph mineral/assemblage, you should be able to make some informed guesses as to the type of parent mineral that was pseudomorphed – see also Table 5.1. Lines of investigation concerning the timing of growth of the parent mineral include:

Did the pseudomorphs grow after a deformation fabric or veining?
Do they contain inclusions, and are these identifiable as minerals?
Are the pseudomorphs less deformed than their host rock?

Table 5.1 *List of common pseudomorph minerals.*

Minerals replaced	Common replacements (most common, in italics)	Less common replacements
Plagioclase Epidote Zeolites (Lawsonite) (Prehnite)	***Epidote group*** Plagioclase Zeolites (sometimes+albite, or a white mica)	Lawsonite Pumpellyite Prehnite (Hydrogrossular in serpentinites)
Olivine	***Serpentine***	Talc Chlorite
Pyroxene	***Amphibole*** Chlorite	
Amphibole Garnet	***Chlorite*** Amphibole	Biotite
Biotite	***Chlorite***	Chlorite+sericite
Kyanite Andalusite Sillimanite	***Sericite*** ***White mica***	Kyanite Andalusite Sillimanite
Feldspar (K or Na) Topaz	–	White mica Sericite
Staurolite Cordierite Chloritoid	***Chlorite+white mica*** ***Chlorite+sericite*** ***Biotite+white mica***	Chlorite
Periclase	***Brucite***	
Blue amphibole Jadeitic pyroxene	***A mat of fibrous green or grey-green amphibole***	

Are the pseudomorphs deformed, or do they define a fabric?
Are pseudomorphs concentrated along veins, contacts, or faults?
Do they occur in reaction zones?
Is it possible to determine the grade of the pseudomorphing minerals?
If inclusions are visible, what is their relationship to the host pseudomorph?

Many pseudomorphed patches have a bulk chemical composition very close to that of the mineral they replace. That is to say, the pseudomorphing involved minimal chemical exchange with the host rock. Compositional similarities can then be used as evidence of the identity of the minerals concerned. Table 5.1 shows useful mineral information for common pseudomorphs. This kind of evidence does not apply in reaction zones or at contacts between siliceous and other rock types. Pseudomorphing there is usually driven by the need to destroy the previous chemical distributions and create new ones.

If a marked compositional zonation exists within the pseudomorph or the margin of its host rock, then the whole texture should be carefully drawn and (as far as possible) annotated with mineral names. If more than a quarter of the pseudomorph is a rim of markedly different material from the core, compositions cannot be reliably used in identification (though identity may still be clear from situation or shape). The causes of zonation may prove difficult to interpret in the field as there are many possibilities (e.g. changes in conditions during growth, original compositional variation, disequilibrium textures, complex porphyroblast–matrix reactions, etc.). Additional field examples of pseudomorphs after metamorphic minerals are shown in Figure 5.13.

5.2.3 Questions and checklist for identified grains and fragments

In the field, the grains and fragments described in this chapter will need to be sketched (e.g. Figures 5.1, 5.2, see also Chapter 2,) to show their shapes, orientations, and distribution among other fabrics of the rock. While recording the information preserved with isolated grains/fragments, you should consider the following checklist (also reproduced in shorter form in Chapter 8 among the other reference tables):

1. Check that annotations convey all the available information.
2. Specifically list any deductions that can be made about origins and grade. State whether there is any local or regional consistency of orientation relationships. These are matters for correlation with other field evidence.

The following list of questions indicates the information needed in sketches, observations, and interpretations of grains and fragments:

Are shapes those of undeformed mineral grains, or the products of deformation? If the latter, are they: (i) ductilely deformed single grains? (ii) pulled-apart, broken single grains? (iii) one of the types with pointed tails? If the last of these, do they consist of one mineral, or do they have a separate resistant core and a deformed rim or tail?

In the case of isolated grains, or of resistant cores, is the mineral metamorphic or pre-metamorphic? If metamorphic, is it of the same grade as the rest of the rock? Is it a pseudomorph of a mineral that could only have existed at different pressure–temperature conditions than those indicated by the main rock mineralogy? Could it be a pseudomorph of a primary (sedimentary or igneous) mineral?

In the case of deformed patches with resistant cores, are the cores single grains, or the broken pieces of earlier single grains? Are the tails granular or bearded? Are the tails composed of minerals with similar or very different composition to the cores? Are the tails composed of minerals found nearby in small veins? Are the cores and tails of a similar metamorphic grade?

Figure 5.13 *Examples of pseudomorphs. (a) Pseudomorphs after lawsonite Syros, Greece. (b) Pseudomorphs after andalusite. QUT teaching sample from the White Blow Formation, east of Mt Isa, Australia (photos Dougal Jerram).*

Are inclusions of small grains visible within porphyroblasts? If so, do they show a preferred orientation of individual grains (a fabric)? Is the orientation of these:

1. A rational one in terms of the crystallographic directions of the host grain?
2. A fabric, or lamination, which the host grain has overgrown during its development?

If the latter, is there a change in orientation at the grain edge? Is such a change a sudden kink, or a curve within the grain margin, or both? Is it 'rolled'?

In the case of grains or augen cores having a markedly non-equidimensional shape, do the shapes have a consistent orientation, or a consistent sense of angular variation from a matrix marker, such as fabric or banding? If so, what are the orientations of grains, fabric, and banding (as appropriate) and asymmetry sense, and is the pattern consistent for one outcrop, a group of outcrops, or for the formation? Does the sense of asymmetry change across major folds?

5.3 Boudins, Shear Pods, and Knockers

Blocks or slabs of rock that have suffered much less strain than their surroundings may be found within deformed and metamorphosed formations, often where significant shear has occurred (e.g. Figure. 5.14). These can have a number of origins and can be described in different ways, depending on their form and relation to the surrounding rock mass. Terms such as boudins, shear pods (e.g. Figure 5.15), and knockers (Figure 5.16) are used to describe some of these types of features. The formation of boudins is indicated in Figure 5.15a, with examples in Figure 5.15b–e. In the case of rotated blocks, boudins, and shear pods, their origins are somewhat clear as their original layers can be traced and reconstructed. The ultimate origin of isolated pods and larger knockers will be less clear, and you might need to search for additional evidence to help piece together their origins.

If the isolated blocks or slabs make up discontinuous but recognizable layers or intrusions, broken up by boudinage or other consequences of shearing, they can be described as part of their host formation: care must still be taken with recording of edges and reactions, as will be noted. If they occur as clearly distinct isolated or scattered masses, they deserve individual attention. In this, you should aim to understand any differences in composition that are associated with variation in the protolith and differences introduced during metamorphic process. You should also try to understand the relationships between different metamorphic histories preserved in parts of the rocks in which strain and grade have been partitioned. A good set of questions to address include:

1. Do *the rock compositions* indicate that the material was of very different origin from the rest of the formation or unit in its pre-deformed state, such as an intrusion into meta-sediments? To help with this, compare compositional categories, for example using Table 3.1, and reference tables in Chapter 8. Pre-existing rock structures such as dikes have different rock properties and potentially different histories, and therefore react differently when deformed. Understanding of this will help in both characterizing the protoliths and determining their deformational history.
2. Does the *location and shape* of bodies suggest that the less deformed mass is of initially more competent composition than its surroundings? Or do they imply that it is different because

Figure 5.14 *Examples of mafic pods. (a) Zoomed-in Greenland pod (photo; see also Figure 4.1). (b) Angular mafic block in Lewisian Gneiss, Scotland (photo (a) courtesy of Susanne Schmid, photo (b) Dougal Jerram).*

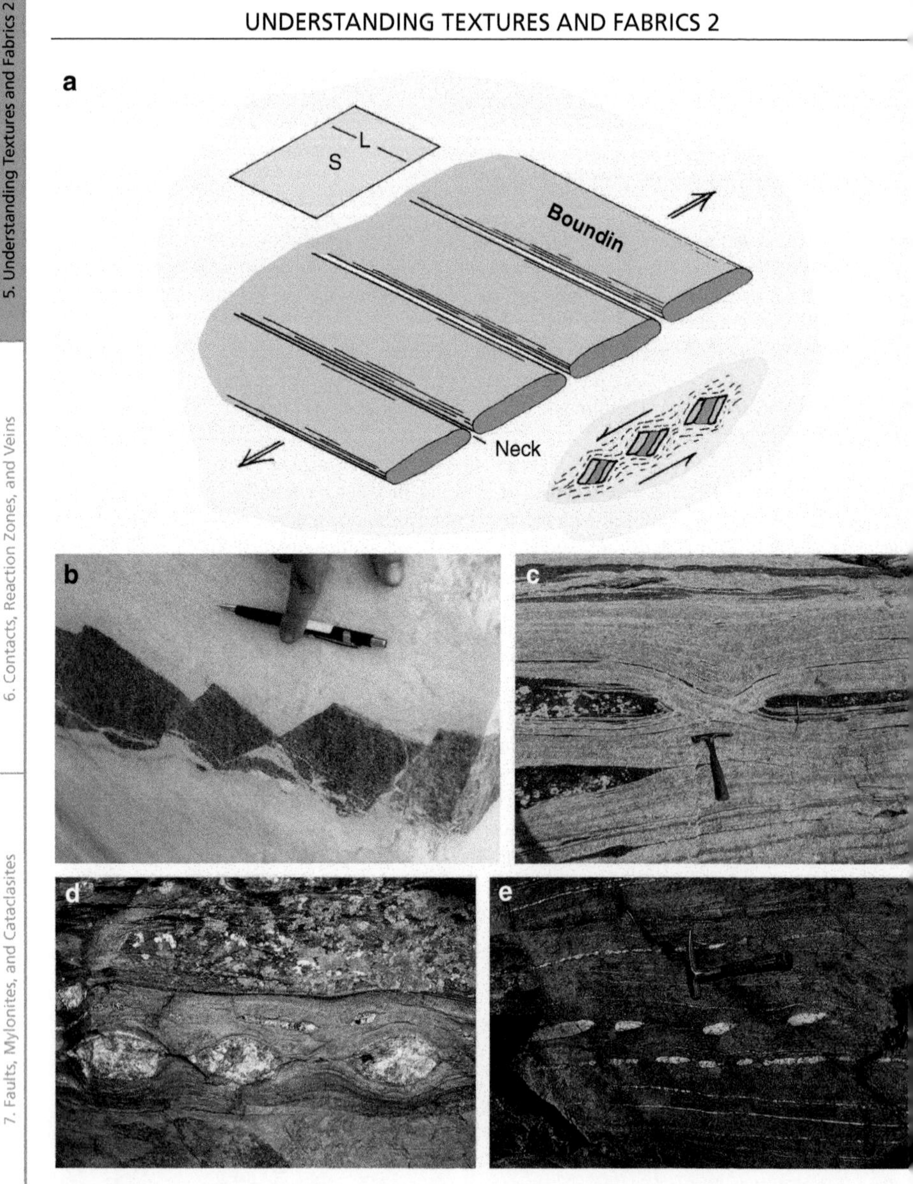

Figure 5.15 Boudins and shear pods. (a) Formation of boudins (Source: redrawn from Best 2003). (b) Boudins in a marble, Naxos, Greece. (c) Amphibolite boudins in gneisses of the Parry Sound domain, Ontario, Canada. (d) Shear pods, Mt. Rushmore, SD, USA. (e) Boudinaged pegmatites, Silver River, Cascades USA (photos: (b) Mark Caddick; (c) Christoph Schrank; (d), and (e) Jim Talbot).

localized structures have produced areas of lower and higher strain from a similar set of materials? A *boudin* or *pulled-apart* block (e.g. Figure 5.15b) is typical of more competent materials surrounded by a host that deformed more readily at the conditions of shearing. A shear pod (a volume bounded by zones of higher shear strain) need not have been more competent than

Figure 5.16 *A knocker of mafic eclogite, Syros, Greece. The eclogite was once surrounded by serpentine and chlorite-rich lithologies that have preferentially eroded (photo Dougal Jerram).*

adjacent rocks of the same composition, but shear pods are commonly bounded by both less competent compositions and by shear zones through material of the same composition (e.g. Figure 5.15d). For example; a fold hinge zone is commonly an area of lower strain than fold limbs of similar composition.

3. Do the *same minerals and similar texture* occur in the less deformed block and its more deformed matrix (suggesting that the processes creating the mineral assemblage have continued to operate throughout areas of different strain)? Or is the less deformed block a remnant, having an earlier mineral assemblage that has been destroyed elsewhere during deformation?
4. Is it possible to discern *different stages of development* within the rock? For example, are the isolated pods always clearly defined, or do some show partial inclusion of themselves within the host rock fabric? In some instances, it is possible to find highly deformed host material with angular, somewhat coherent, pods alongside areas in which the pods are streaking out into the hosts (e.g. Figure 5.17).

In the field, the first notes about a less-deformed block within a sheared outcrop should address these questions, and should include a sketch of the shape of the block and the traces of any banding and fabric through and around it. If more than one block of similar type exists, a note should be made about whether there is any consistent relationship for the set, particularly of the 'sense' of angular discordance between the fabric, banding, and the boundaries of shear pods. You may, for example, be able to find that boudins are developed in one part of a metamorphic terrain and show a clear gradation towards more distinctly isolated units in regions containing more highly deformed and sheared units. In this case, you have found evidence for how the same protolith will deform under different conditions.

Following this, notes should give evidence of whether the less deformed rocks provide a new and as yet undescribed composition, texture, fabric, or set of small structures. If these show a different grade of

Figure 5.17 *Examples of mafic pods and mafic layers within the Lewisian complex, Achmelvich, Scotland. The mafic material, which was probably originally intrusions within a granite host, appears as streaky layers and discontinuous pulled-out arrays of pods. (a) Angular partly rotated bock; (b) more ductile and sheared mafic material; (c) ductile sheared and rotated example (photos Dougal Jerram).*

metamorphism to previously described outcrops, or display strikingly different structural features, they may play an important role in establishing the geological history of the outcrop of region.

Finally, relationships at the block's edge should be thoroughly described. Is the contact sharp or diffuse? Banding within the block may pass into the matrix, or may be cut off by it. Some blocks may be surrounded by a reaction-zone with their host (see Section 6.3, Reaction zones and chemical changes at contacts). Occasionally, some bands trace into the matrix while others do not. This may reflect competence contrasts, or may result from thinning down of a band bounded by reaction zones to the extent that only the reaction zone exists in the highly deformed matrix.

Keep in mind when describing these features that any banding or compositional variation can have one of multiple potential origins, including:

1. Pre-metamorphic in origin. For example, bedding, contact relationships, and even fossils (though in most cases these features may be altered beyond normal recognition).
2. Metamorphic. Relating to local mineral changes.
3. Metasomatic. Involving the chemical transport and mineral change associated with fluids.
4. Structural. Relating to the rock deformation.

CONTACTS, REACTION ZONES, AND VEINS

Complex contact between granite and metamorphosed banded country rock at the edge of the Ross of Mull granite pluton, on the Isle of Mull, NW Scotland (photo Dougal Jerram).

6

CONTACTS, REACTION ZONES, AND VEINS

Contact metamorphism between igneous rocks and their hosts represents a style of metamorphism that focuses into relatively small zones and can occur in shallow and surficial conditions as well as deep within metamorphic terrains. Veins are a common phenomenon and are often associated with igneous (hydrothermal) systems, but can also be formed by fluids derived from within metamorphosing rock sequences experiencing regional metamorphism. The distribution of reaction zones are also somewhat constrained by the occurrence of bodies of different composition and nature, but the detail of reaction zone and metasomatic alteration may be somewhat more complex than simple contact metamorphism. Where rock compositions vary at sharp contacts, these can often be zones in which complex sequences of chemical reactions occur during metamorphism.

In this chapter we shall firstly consider intrusive contacts and their associated metamorphism, with examples of different levels of aureole development. We then discuss 'veins' in the broadest sense, purposely making no distinction in treatment between those of known igneous origin and those of unknown or of non-igneous origins. Finally, we discuss reaction zones and chemical changes at contacts.

6.1 Igneous Contacts – Aureoles and Metasomatism

Contacts between igneous rocks and others can be complicated. Their origins may be depositional, intrusive, or faulted. If igneous material was deposited or intruded hot, the adjacent country rocks may have experienced a series of heating effects. Magma or lava may also have been affected by the proximity of a cool, static, country rock. Thereafter, the contact is liable to form a sequence of reaction-zone effects as the complex cools. It then becomes subject to all the same normal reactions as any other contact (Section 6.2).

6.1.1 Intrusion geometry

When addressing contact metamorphosed rocks, the shape of the igneous body and the contact should be carefully considered. The complexity of the contact can vary significantly, depending on the type and scale of intrusion, and the way it has reacted/interacted with the country rock (Figure 6.1a). Small intrusions may have the same or similar forms and patterns as veins (see Section 6.2). Additional details may include small side-intrusions, back-intrusions of mobilized country rock, and the effects of deformation during or after igneous consolidation (e.g. Figure 6.1b).

In the field, intrusion shapes should be mapped out or sketched depending on size. Details of contacts should be sketched and annotated with mineral contents and proportions, which often change progressively in thin intrusive fingers of one rock type into another. In many cases, the original geometries and cross cutting relationships between different phases of intrusive events (e.g. cross-cutting dykes) will still be apparent even through significant metamorphism, which will allow the original sequence of emplacement events to be inferred (e.g. Figure 6.1a).

The Field Description of Metamorphic Rocks, Second Edition. Dougal Jerram and Mark Caddick.

Figure 6.1 *Examples of intrusions within a metamorphic context. (a) Complex intrusive contacts with host metasediments, Ross of Mull, NW Scotland. (b) Palaeoproterozoic (~2.1 Ga) deformed/ foliated trondhjemite at Itutinga, Minas Gerais State, Brazil, with leucossome veins (photos: (a) Dougal Jerram; (b) courtesy of Claudio De Morisson Valeriano).*

6.1.2 Contact metamorphic aureoles

When you are dealing with examples of hot igneous rocks intruded into cooler host rocks (be they sedimentary, pre-existing metamorphic, or igneous), there are often mineralogical or textural indications of an aureole around the igneous body in which heat lost from the cooling magma drove recrystallisation in the host. The direct effect of heat on the host rocks, generally termed the 'country rock', typically manifests as either a localised baked contact of a few metres' width or less, or an aureole that can be several 10s to 100s of meters or even km-scale. These rocks have experienced contact metamorphism, that is recrystallisation of existing phases or the growth of new minerals at higher temperature, as a result of this heat. The thickness of the contact zone varies with the size of the intrusion, the rate and depth of its emplacement, the ambient country rock temperature before intrusion, the temperature of the magma being emplaced, the composition of the country rock, and the availability of hydrous fluids. As a general rule, contact aureoles are better developed adjacent to shallow intrusions, very thick intrusive bodies, at the base of thick lava flows, and in systems with availability of magmatic, meteoric, or metamorphic fluids. Typical effects of contact heating include:

1. Bleaching of quartzitic rocks (and colour changes more generally through the contact zones, e.g. Figure 6.2).
2. You may be able to discern a zone of hardening (induration) along the contact areas (e.g. Figure 6.2). This is evidence of heat transfer into the host rock resulting in mineralogical changes (e.g. precipitation in pore spaces of shallow sediments).
3. Development of spots in certain rocks (e.g. Figures 6.2 and 6.3).
4. Hornfels formation from fine-grained rock-types (Table 3.3 and Figure 6.3).
5. Blurring of the edges of previous large mineral grains or patches (including amygdales, phenocrysts and single-mineral clasts), and of the textures of coarse crystalline rocks.
6. In extreme cases, some partial melting of country rock may happen (though this may only be apparent by later inspection in thin section). In this case, significant heat was added to the country rock to elevate temperature above its solidus. This can be achieved by long-term magma flux through feeder intrusions and by the emplacement of particularly hot magmas, such as primitive mafic melts that would crystallise to form basalt, dolerite/diabase, or gabbro (e.g. Figure 6.4, see also Figure 8.3).

In the field, note what kind of features are diagnostic of the contact metamorphism, and how they change on a traverse towards the igneous contact, starting from beyond the aureole (i.e. in an area showing no apparent effect of contact heating). If possible, a sequence of aureole zones should be defined using the simplest characters available. Such characters are colour, grain size, spacing of joints, type of fracture when hit by a hammer, or the minerals in veins (e.g. prehnite then epidote in basic or intermediate igneous country rocks). Any identifiable changes in minerals in the host rocks should be recorded, but these rocks are typically fine-grained so it should be no surprise if this can only be accomplished with the aid of thin sections made from collected samples. In this case, simple features such as the way the rocks weather may help indicate the area of change, which may be subtle or unclear (Figure 6.5). Aureole zones should be mapped, where big enough and if exposure is adequate. This may mean making a more detailed local map around an intrusion, if the general map is of unsuitable scale. Either in addition to, or in place of, a map in very poorly exposed ground, an example transect should be constructed along a line of good exposure that traverses the aureole. This should record zone widths and variations within zones that are not part of the general zone definitions. Good use of annotated photographs can help to locate and capture the finer details at high resolution as well (e.g. Chapter 8, Figure 8.3). Remember that the types of contact metamorphic zone will depend on the host rock composition and the degree of metamorphism (see Figure 3.7).

6.1.3 Aureoles in regional metamorphic terrains

In many cases, the timing of magma emplacement in relation to regional metamorphism is important, particularly in cases such as granite genesis and emplacement at specific times during the development of mountain building events. In most examples there is likely to be some discrepancy between the timing of intrusion and the timing of regional metamorphism, and you should note that

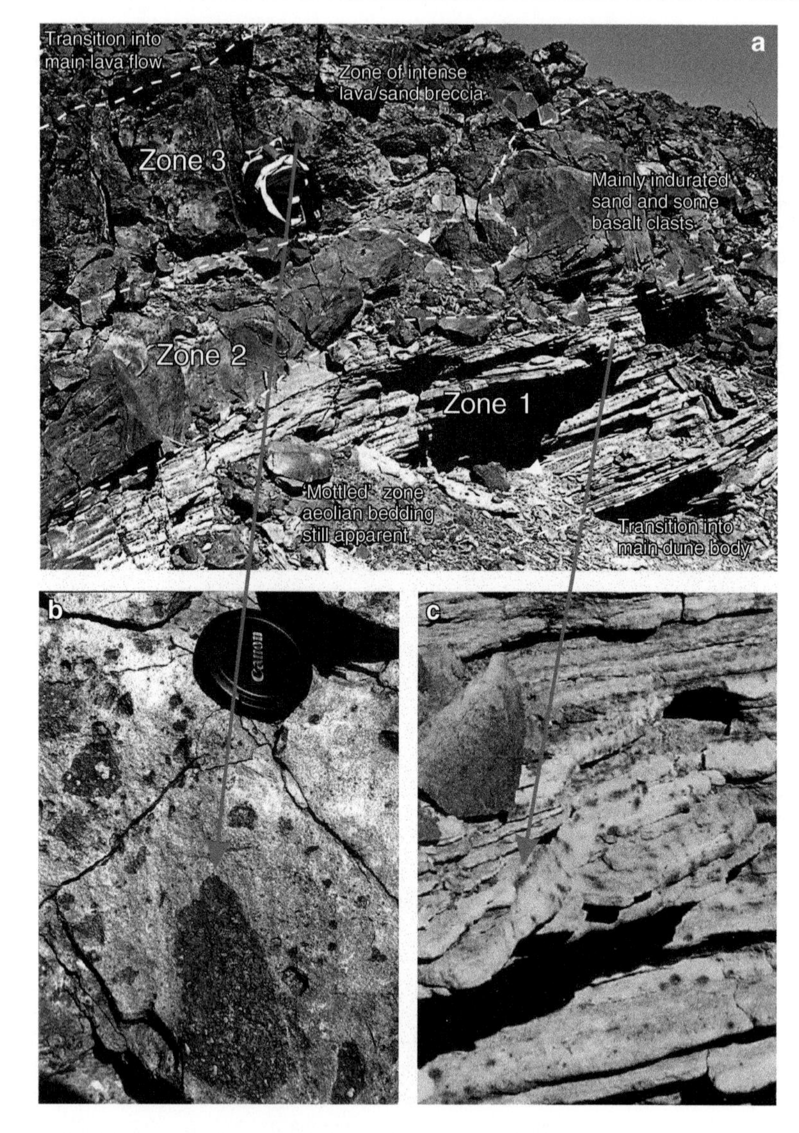

Figure 6.2 *(a) Example of an aureole beneath a large ponded lava flow on top of aeolian sands. A more intense zone of sediment lava breccia with 'fused' bleached sandstone (zone 3, e.g. 'b') is followed by an indurated brown-coloured highly cemented hard zone (zone 2). The sands below this can be seen to be partially bleached and spotted with clusters of iron oxides (zone 1, e.g. 'c'), which then transitions into the main sand body (photos Dougal Jerram).*

the characteristic timescales of each are different, with magma emplacement and cooling typically taking a significantly shorter time than a complete orogenic event.

Igneous intrusion into deep metamorphic rocks, already at moderate temperature and grade, produces a broad 'aureole' of increased metamorphic grade, which is normally only defined by minerals,

Figure 6.3 *Examples of hornfels. (a) Cordierite porphyroblasts in a calc schist, Namibia. (b) Chiastolite porphyroblasts in slate, Lake District, UK (photos Dougal Jerram).*

rather than by obvious changes of physical properties, and which may not perfectly follow the outlines of the igneous contact. As such, it is easily missed without careful, general mapping (if it can be distinguished at all from variations in the regional metamorphism).

Intrusion of igneous rocks into regionally metamorphosed rocks after they have been exhumed to a shallow depth will superimpose a definable aureole onto pre-existing schists, etc. Hornfelses, visibly derived from medium- or high-grade schists, should be discernible at outcrop. Their loss of schistosity and tendency to irregular curved fracture may be taken as evidence of uplift after regional metamorphism and before intrusion. If the earlier metamorphism was at very high grade, however, it is possibly that the rocks may show little to no obvious effect of the contact event.

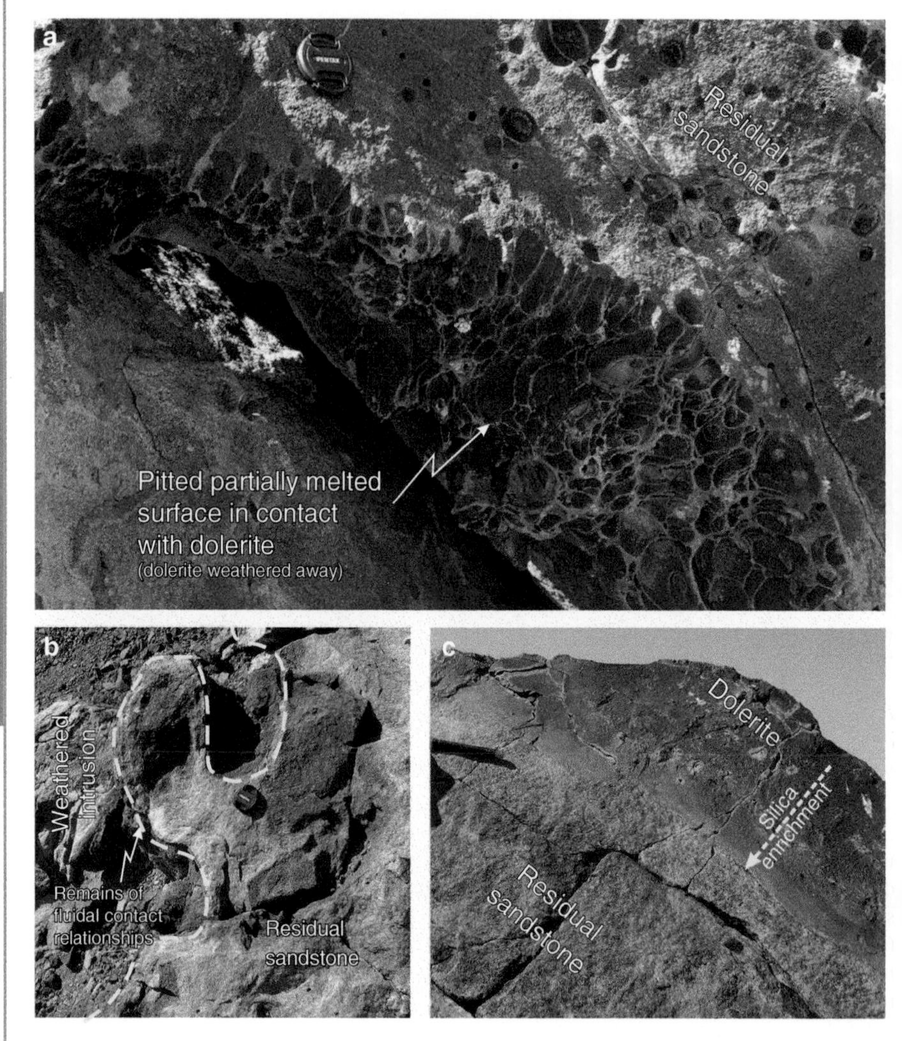

Figure 6.4 *Examples of melting at contacts. (a) Pitted sandstone contact against 'now eroded' dolerite intrusion (morphology of contact similar to regmaglypts seen on surfaces of meteorites). (b) Fluidal geometry preserved at contact of sediment/intrusion. (c) Silica melt from sandstone grading into dolerite intrusion at contact (photos/interpretation courtesy of Clayton Grove).*

The aureole of an intrusion that later experienced regional metamorphism is unlikely to be recognizable, unless the regional event is of low or very low grade. In this case, the hornfelsed rocks will generally constitute a more competent mass than their un-hornfelsed equivalents, and this will be shown up by the structural relationships of a later fabric, and possibly by a difference in the type of cleavage. You may also find some of the original hornfels minerals as pseudomorphs (replaced by new metamorphic minerals that formed at the regional grade).

Figure 6.5 *Subtle effected aureoles around a picritic dyke in Namibia. The zones either side of the dyke are best picked out in the field by the weathering, where they are more prominent. Careful observation can show how the zones thin as the dykes thin (photo Dougal Jerram).*

A classic example of an intrusion in a regional metamorphic setting is the Ross of Mull Granite in NW Scotland (Figure 6.6). Here, regionally metamorphosed rocks (kyanite bearing), were heated during intrusion of the granite, causing kyanite to be pseudomorphed by andalusite and eventually sillimanite close to the contact (e.g. Figure 6.7). The phase diagram shown as an inset in Figure 6.6 (see also Chapter 3) hints at the different pressure and conditions of the regional and contact metamorphism as the conditions are moved along the alumino-silicate *P–T* diagram. Other metamorphic reactions involving cordierite, garnet, and micas also occurred in these rocks, but the large grain size of the aluminium silicates provides a good marker to see some of these changes in the field, and with the use of the alumino-silicate *P–T* diagram (Figure 6.6) it acts a good way to visualise the influence of the Ross of Mull pluton. In each case where a large igneous body has imposed an aureole on a regional metamorphic assemblage, the composition of the protolith, and any variations within it, will control the elements available for the growth of new minerals that may indicate pressure and temperature of contact heating (e.g., Figure 3.7). As such, it is important to search for any variations in the host that may lend to better production of key indicator minerals.

6.1.4 Metasomatism

This name is given to chemical alteration of country rocks, with the assumption (which is now considered a significant simplification) that:

1. Metasomatising fluids emanated primarily from within the igneous intrusion.
2. Metasomatism occurred while the magmatic mass was hot.

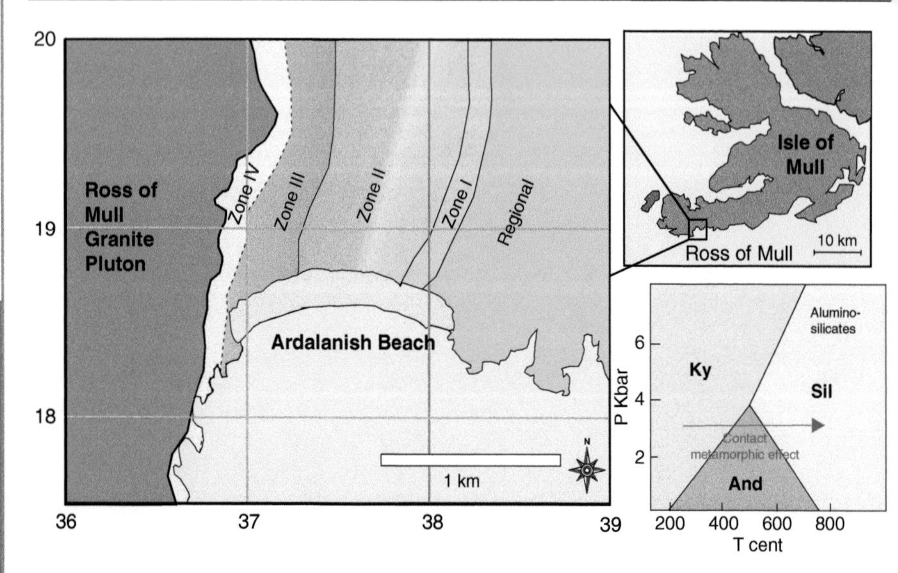

Figure 6.6 *Simplified map of Ross of Mull contact metamorphic zones found around the Ross of Mull Granite pluton on the Isle of Mull, Scotland (UK Ordnance survey grid). The inset highlights the effect of the contact metamorphism on the aluminosilicates in the rocks (see figure 6.7 for examples).*

More broadly, metasomatism can be driven by the influx of fluids produced by any process, and the interaction of those fluids (and any components in solution in them) with the rocks they encounter. Such alterations are highlighted by a loss of some minerals from the normal assemblage of the host rock, often accompanied by the addition of a 'metasomatic' mineral rich in a particular minor element (e.g. tourmaline, rich in boron). Tourmaline, fluorite, and topaz are commonly metasomatic, but there are many other possible minerals, including a number of pale-coloured 'white micas'.

In the field, mineral occurrences may be declared metasomatic if they are rich in 'metasomatic' minerals, if chemical alterations are strongly localised along zones or around veins near an igneous contact, or if the number of minerals present in the rock has been reduced. In extreme cases, metasomatism can drive rocks towards being monomineralic. Figure 6.8, an example of dykes cutting through carbonates, highlights some of the complex relationships that are possible. Here, reactant metasomatic fluids likely reacted with the carbonate to form striped dark and light coloured bands along the sides of the dyke or as apophyses from them. In such an example, the processes may be further complicated by the fact that the host rock is more reactive to high-temperature aqueous fluids than less 'reactive' siliciclastic hosts might be. Sometimes, such zones can be mapped out, and they may even show complex internal zoning revealed by variations in mineral contents and types, and in textures. This should be done where possible. Otherwise, localities should be marked on the map, and the metasomatic rocks of each locality described. Overall, it is important to produce:

1. Sketches of the minerals and textures of the most extreme metasomatic rocks.
2. An ordered list of each country rock's minerals in their order of replacement by the metasomatic mineral (e.g. tourmaline replaces: hornblende, biotite, plagioclase, K-feldspar, quartz).

Figure 6.7 *Examples of metamorphic textures from the Ross of Mull aureole, Scotland. (a and b) Blades of blue kyanite: partially transformed to andalusite (pink) from Dun Fuinn within zone II. (c) Large blades of kyanite, now pseudomorphed by andalusite, on Ardalanish Beach in zone III. (d) Spectacular aluminosilicate knot within zone IV, mostly white sillimanite but also contains small grains of blue kyanite and pink andalusite. The knot is surrounded by thin biotite rich rim with small garnet grains, with the knot and rim surrounded by granular leucosome (photos courtesy of Dave Prior).*

6.1.5 The margin of the igneous body

Three causes of compositional variations in the margins of intrusions need to be distinguished.

1. Magmatic variations. These may be the product either of more than one magma, if the intrusion is composite, or of crystal sorting. Sorting may be by grain size or mineral species, and may result from layering at the base of a magma chamber or from shearing of a magma boundary

Figure 6.8 *Complex metasomatic zones around the edges of multiple dyke sheets into marbles, Isle of Skye, Scotland, UK (photos Dougal Jerram).*

layer next to a contact, amongst other factors. There are also the possibilities of crystalline growth outwards from the wall of an intrusion, and of fine-grained chilled margins. All these variations are originally in the size and proportions of igneous minerals. An additional complication is the concentration of late stage, volatile rich, melt within the crystallising magma, which can separate and form veins and pegmatites. These features, although originating with the intrusion, may be hard to distinguish when significant post intrusion metamorphism has occurred.

2. Effects of hot circulating fluids. These effects may be common, but are only likely to be discernible in the field if they involve changes that could not have occurred after cooling of the intrusion. The two most likely cases are the late-stage growth of hornblende in mafic intrusions, and metasomatic alterations to the intrusive igneous rock itself (as distinct from the country rock).
3. Back-reaction. Reaction-zone effects are usually restricted to a few metres, or sometimes centimetres, from the contact. They vary with the type of country rock. If the grade of reaction-zone minerals is greater than that of the country rock, reaction must have occurred before cooling of the intrusion. These effects involve major changes in minerals, sometimes in a series of zones, along the contact. General considerations applying to reaction zones are discussed in Section 6.3.

In the field, variations in minerals near the margin of an intrusion should be searched for and categorized on the basis of mineral type, textures, and distributions. Textures should be drawn, paying particular care to highlight the variation of texture with location over distances of a few metres.

6.2 Veins and Pegmatites

Veins and other small intrusions (collectively called veins in this section) may make up only a small proportion of a rock unit, and give the appearance of being of extraneous origin (brought into the unit from elsewhere), but they can be full of useful information. They may:

1. Have been derived totally from out of the surrounding rock.
2. Have been brought in from great distances along fractures.

3. Have formed from fluids released by metamorphic reactions in nearby but compositionally different rocks.
4. Result from equilibration of a foreign fluid with the local host rock, with partial transfer of material before crystallization.

It may not be possible in the field to distinguish which of these possibilities is most likely to have occurred. It may also not be clear whether the vein crystallised from an aqueous (or other non-silicate) fluid, or whether it is a crystallised melt (particularly if it has suffered later deformation). A rock-mass description must record veins, but clear distinction must be made between observations and interpretations, where inferences are possible.

6.2.1 Vein patterns

Veins are commonly concentrated in bands, and their style can be helpful in determining rock unit heterogeneities, and deformation (e.g. Chapter 4, Section 4.3), grading and shear directions. Extension of competent rock bands can result in cross-veins between the pulled-apart sections. These initially develop at a high angle to the band, but may have been deformed later. They can highlight discrete bands, grading, and otherwise ill-defined layers in banded rocks (e.g. Figure 6.9).

In shear zones, extension veins initiate at about 45° to the zone boundaries, so that en-echelon arrays are developed (Figure 6.10 and Figure 4.18 in Chapter 4). These arrays highlight the role of shear where it might otherwise go unnoticed, particularly if strain is very low. In some cases, a shear zone is a rock band of below-average competence in a stratified sequence. Vein orientation and comparison of rock strains should distinguish between an extended competent layer and a sheared incompetent one. In other cases, shear zones do not correspond to rock bands, and may develop in conjugate sets (Figure 6.11). These usually cut locally more competent units. Their individual veins may vary greatly in aspect ratio, and may be straight or sigmoidal (Figures 6.10–6.12). They lie approximately perpendicular to any local shear-zone fabric, having opposite sense to the shear direction, but can be deformed into the shear direction. Shear zones of veins can exist on any scale from microscopic to hundreds of metres in width.

Figure 6.9 *Compositional banding (bedding) where quartz veins are more strongly developed in some beds, from Ticino, Switzerland (photo Mark Caddick).*

Figure 6.10 *En-echelon veins/tension gashes rotated during a progressive shear. Millook Haven, Cornwall (photo courtesy of Jim Talbot).*

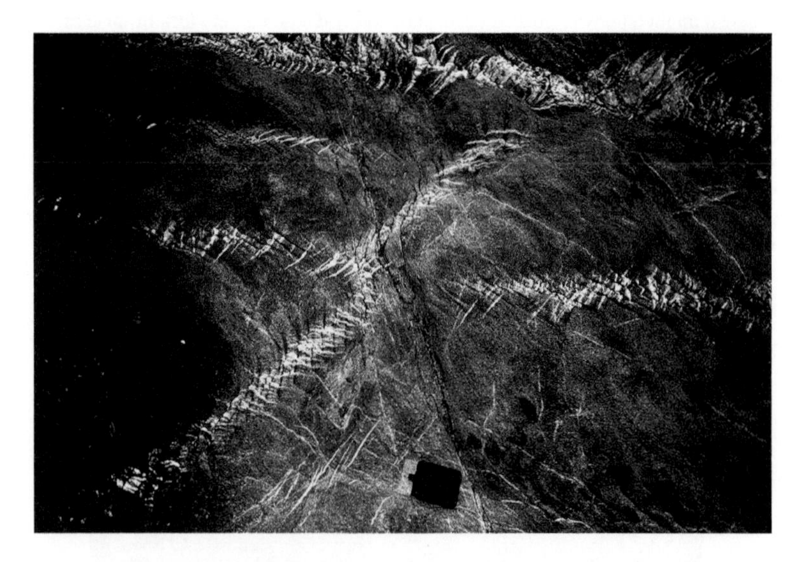

Figure 6.11 *Conjugate en-echelon tension gashes in Old Red Sandstone. Marloes sands St. Bride's, Dyfed, Wales (photo courtesy of Jim Talbot).*

6.2.2 Vein shapes, orientations, and fibres

Wherever sheet veins or intrusions are found, they may have step-like offsets or specific geometries that should be identified and sketched (e.g. Figure 6.12). Shapes to look out for at vein ends include splays with a threshold take-off angle, and veins that pass into a radiating brush-like array of microscopic veinlets (usually discernible only by weathering). Any irregularity of opposing vein margins may match across the vein, showing the general opening direction, provided this lies approximately

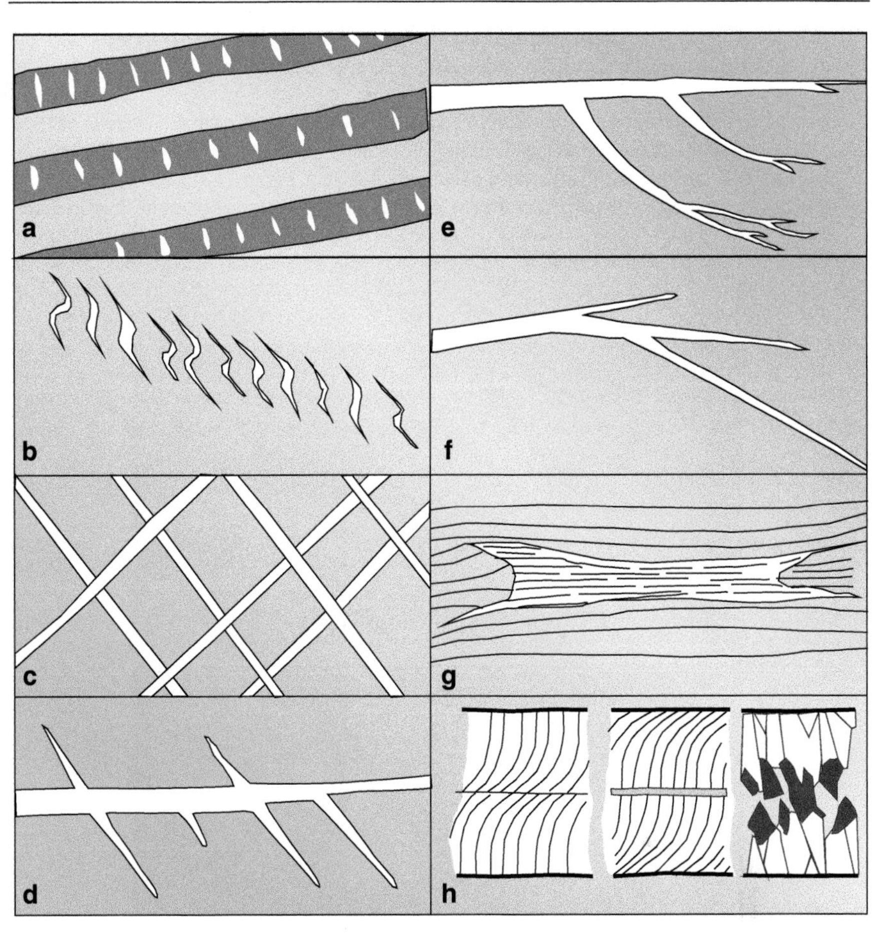

Figure 6.12 *Vein patterns, shapes, and elements. (a) Extensional veins perpendicular to their bands, which have been lengthened. (b) En-echelon extensional veins along shear zones. (c) Veins intruding conjugate shear fractures, which may have initiated earlier. (d) Vein off-shoots picking out structure or fabric of the host rock. (e) Vein splays near a termination. (f) Rotation of extension direction between successive propagation of branches of an extension vein. (g) Vein fibres (centre) along the length of a vein lying parallel to cleavage (trace indicated by lines around the vein). (A cross-vein with its original width extended to become far longer than its original length.) (h) Rotation of extension direction shown by syntaxial fibres (left) and antitaxial fibres (centre). Different mineral generations (right).*

within the plane of the exposure surface. Where banding is cut by veins, the sense of offset of its trace across a vein may be visible on a rock surface, but it may need recording on several surfaces to show the true opening direction (which will otherwise remain ambiguous).

Vein shapes may result from post-veining deformation. In particular, veins may be the active element in folding (Chapter 4, Section 4.6) or boudinage. Where strain is high, veins may become unidentifiable and effectively concordant with other lithological elements, and so contribute to banding and to lamination fabrics. Any intermediate deformation stage which suggests this should be recorded.

Veins that continue to develop during deformation may show relationships between their shapes and those of constituent mineral grains, if the veins widen by successive extension of their grains into fibres.

1. Continuous vein fibres can join the points on opposing margins which were originally in contact.
2. If the opening direction has changed during the history of the vein's opening, this may be recorded by curved fibres, whether syntaxial or antitaxial (e.g. Figures 6.12h and 6.13).
3. In some instances, initial veins of a limited original length may be extended during further deformation. Fibres within veins can show whether they have opened across or along their length.
4. Further metamorphism can cause new mineral phases to grow across the original veins (e.g. Figure 6.13c).

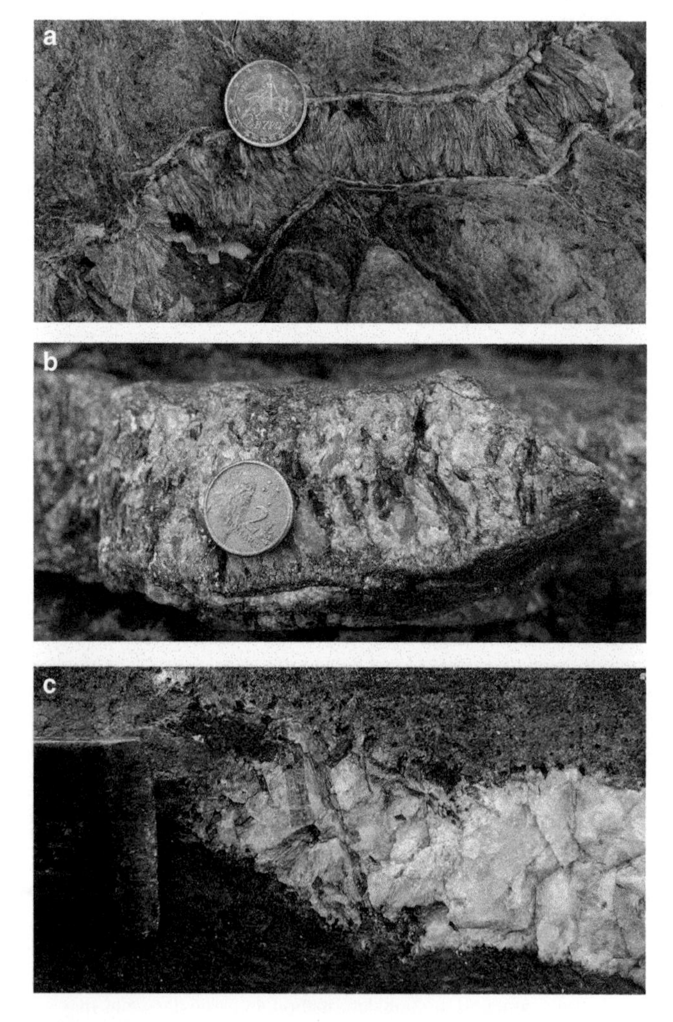

Figure 6.13 *(a) Tremolite and chlorite growing into a vein, Syros, Greece. (b) Elongate-blocky syntaxial lepidolite-bearing quartz vein, Forsayth, Queensland. (c) Kyanite crystals growing across a quartz vein, Tauern Window, Austria (photos: (b) courtesy of Christoph Schrank; (a and c) Mark Caddick).*

6.2.3 Recording veins

In the field, record first any features similar to those mentioned or illustrated above. That is:

1. *The pattern of veins*, the orientation of this pattern, its size, and what it represents.
2. *Individual vein shapes*, dimensions, and orientations.
3. *Opening directions*, and the type of evidence that reveals this.

Annotated sketches are the best way to do this.

In structurally complex rocks, try to answer the following questions:

- Are the veins folded (e.g. Figure 6.14)? Do veins cross-cut folds, or lie within fold structures in some rational orientation or position that shows the folds to be earlier?

Figure 6.14 *Deformation of vein structures. (a) Folded veins cross-cut by later veins, Welcombe Mouth, N. of Bude, Devon. (b) Folding and flattening of pre-formed extension fractures, Jæren Nappe Complex, Vigdel Fort, Norway (photos: (a) courtesy of Jim Talbot; (b) courtesy of Tonje Lund).*

- Are veins and banding oblique to each other, and are both folded? If so, which is the active feature of the folding?
- Is there shearing along the vein? If so, is there evidence to show whether the shear developed because of the existence of the vein, or whether the vein was located by the existence of the shear-zone?
- Is strain concentrated at one side of the vein, or is it symmetrical? Is there concordance of any deformation fabric in the vein with a fabric outside (taking into account refraction, which is very likely to occur due to competence contrasts with the host rock)?
- Are there more complex geometrical relationships between vein and deformation?

Then, draw an example of grain shapes and textures within the veins. They may be drusy, fibrous, comb-structured, etc., but should not be represented merely by such words. Annotate the sketch with the names of minerals (or brief descriptions if minerals cannot be identified) and then answer these further questions about minerals of the vein and its host rocks:

- What are the proportions of minerals? Can the same minerals be identified in the host rock?
- Do mineral proportions vary:
  1. From the centre to the edge, gradationally, cyclically, or step-wise?
  2. Along the vein? If so, does this show any correlation with the local host rock type?
  3. As several slightly cross-cutting bands, representing several episodes of veining?
- Are there early minerals (or relics) in the vein, which may once have existed in the host rock also? (Draw shapes.)
- Are there minerals in the vein that only occur as patchy late growths in the host rock?
- Are there accessory minerals, of potential geochemical or economic interest, concentrated in the vein, or along its margins?
- Does the host rock change on approaching the vein? If so:
  1. Does the rock show gradual or successive sudden changes of mineral content on approaching the vein, suggesting a chemical reaction between the two? If so, what are the changes discernible within that portion which was host rock? Is there evidence of complementary alteration of the vein edge? Is the alteration due to fluids associated with emplacement of the vein (either reaction between those fluids and the wall rock, or those fluids acting as a catalyst for reactions, see Figure 6.15)? Is there any separate reaction zone of new material separating the two? (See also Section 6.3).
  2. Does the host rock develop an assemblage of different metamorphic grade at its margin (see tables in Chapter 8)?
  3. If the change is only one of mineral proportions, does the rock become more like the vein in mineral proportions nearer the contact, or does it become depleted in the vein minerals (suggesting that the vein derived, at least in part, from the local host rock)?

6.3 Reaction Zones and Chemical Changes at Contacts

Chemical reaction between adjacent rocks subjected to metamorphism is both common and commonly overlooked. Very abnormal compositions, which are easily identified, are usually restricted to a few metres, or less, from contacts (Figure 6.16a). Less easily visible changes may extend further. As these changes are likely to be largely controlled by fluids, and contacts can provide preferred pathways for fluid movement, it may be difficult to distinguish some reaction effects from hydrothermal ones. The processes can be complementary.

Figure 6.15 *Reaction at wall of vein in a garnet blueschist, Syros, Greece (photo Mark Caddick).*

6.3.1 Recognition of reaction

In the field, the first task is to recognize reaction. In simple cases this is shown either by the loss of one or more minerals from the assemblage that is prevalent further from the contact, or by a change in composition of one or more minerals. Changes in mineral composition are possibly visible as changes in colour, but can often only be ascertained after more detailed laboratory study. More locally at the contact, extreme differences in the chemical potentials of components between the two rock types can drive the formation of completely new minerals.

Whether or not mineral changes are visible, a record should be made of any general features that suggest that adjacent rocks have affected each other. Bleaching, blackening, or surface staining of the rock, or different resistance to erosion along a contact all suggest reaction. These should be reported, as should any evidence for possible alternative causes of alteration, such as localised cataclasis (i.e. fault-related changes, see Chapter 7).

6.3.2 Reaction zones

Where extreme reaction and/or chemical change has occurred, you may find multiple bands of reaction-zone rocks. At original compositional interfaces you may expect to find bands of rock that lack any record of their previous (pre-reaction) structure, texture, or mineralogy. This may be entirely new material, precipitated during the reaction, but it may also contain earlier rocks that have been so altered as to destroy any trace of their original character (e.g. Figure 6.16b). This zone might

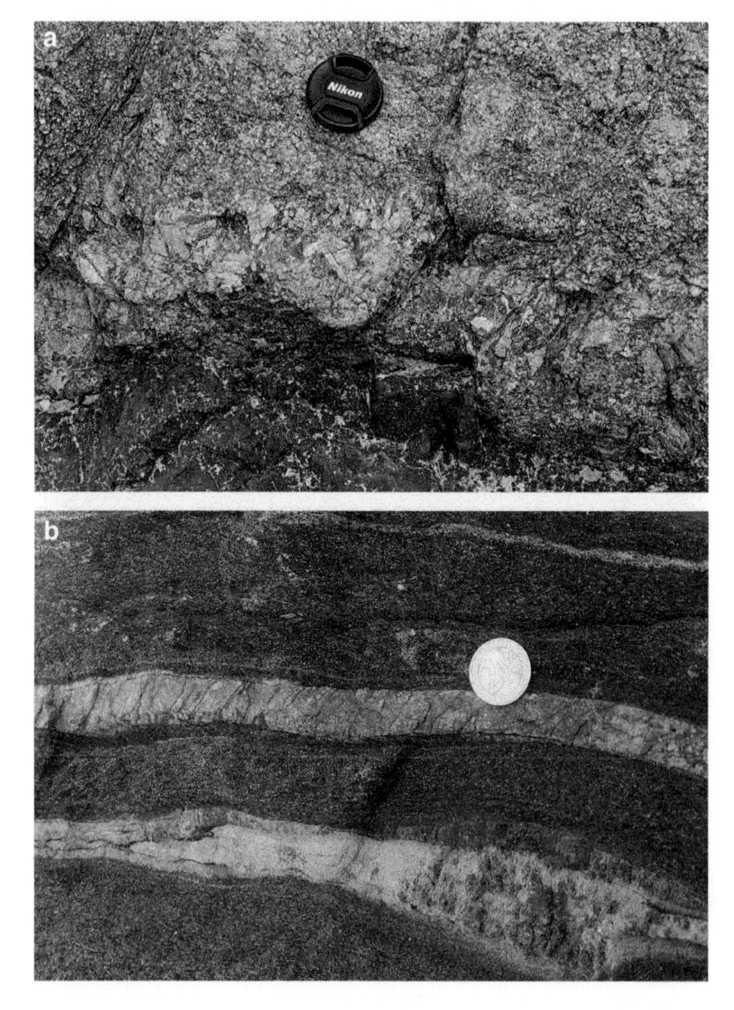

Figure 6.16 *Examples of reaction zones and contacts. (a) Pegmatitic vein parallel to the contact between a Palaeoproterozoic orthogneisses and a Mesozoic dolerite, with the vein now containing an epidote (green)/K-feldspar texture – Praia das Conchas, Cabo Frio, RJ, SE Brazil. (b) Amphibole-rich zones (dark bands) flanking spessartine-rich garnet beds (pale pink) from Syros, Greece (photos: (a) courtesy of Isabela Carmo; (b) courtesy of Mark Caddick).*

be flanked on either side by rocks which, despite chemical alteration, retain features of their previous rock types.

Outside these zones may be others that are only modestly altered, grading towards completely unaltered wall rocks. Any new mineral in these zones is likely to be seen first as a rare alteration, then possibly as a pseudomorphic replacement (depending on the availability of chemically appropriate precursor minerals). Close to the contact, the new mineral may successively replace more and more of the earlier assemblage. Such replacements are often well displayed by metasomatic minerals and by back-reaction replacements of igneous textures at intrusive contacts (Sections 6.1.4 and 6.1.5).

Figure 6.17 *Exceptional mm scale mineral reaction zones of (centre out) orthopyroxene (OPX)-clinopyroxene (CPX)-garnet (GT) in a plagioclase anorthosite (An), Holsnøy, Bergen Arcs, Norway (photo courtesy of Chris Clark; see also Figure 3.14 with scale).*

In the field, the textures of different stages of alteration and metamorphic reaction should be sketched, and a record made of whether they define a regular series of zones, or whether replacement increases patchily towards joints, compositional bands, or foliation planes that might have been channels for fluid transport. It should be remembered that, while reaction 'zones' are often seen within the rock mass, they can be concentrated around very discrete areas in which an extreme chemical variation exists over a small scale (e.g. Figure 6.17; see also Chapter 3 Figure 3.14). If zones are regular, their widths should be measured, perhaps as a transect in larger bands. Otherwise, an example of their irregularity should be sketched or grid-mapped. If several new minerals appear as the contact is approached, their order should be recorded in different places and particularly in different original rock types. The consistency of their order, or of correlation between order and original rock type, should be recorded.

Within the central zone of new material, relationships are usually simple. The number of minerals is normally less than outside, and changes are usually regular (unless complicated by later deformation). If several bands occur, having the same few minerals and similar texture, they are likely to be successive generations, and may have slight cross-cutting geometry. Only with the most extreme differences in rock type are there likely to be several zones having completely different mineral assemblages (the best known examples of which are at contacts of either ultramafic or carbonate rocks with siliceous rocks). Straightforward reaction between adjacent masses produces an unreversed simple sequence of zones each of fewer minerals than the rocks at the sides. Any case of very many bands, of reversals of the sequence, or of irregular accumulations of a number of minerals is dominantly of vein or hydrothermal origin, not a reaction-zone.

In the field, the minerals, textures, and zone widths should be recorded. An annotated sketch is best. It is more important to collect and report this evidence than to decide between reaction zone and another kind of deposit if this is in doubt.

6.3.3 Deformation of reaction zones

Reaction zones are bands of rock that differ greatly in their physical properties, the banding constituting a general anisotropy in between two adjacent masses, which probably differ in competence.

5. Understanding Textures and Fabrics 2
6. Contacts, Reaction Zones, and Veins
7. Faults, Mylonites, and Cataclasites
8. Summary Tables, Checklists, and Mapping Report Advice

In any subsequent deformation, reaction zones are liable to be strongly disrupted, producing features such as isoclinal folds, boudins, and blocks of monomineralic or bimineralic material, lobed contacts, and 'tectonic intrusions'. In the field, look out for disrupted zones of a few simple materials, and draw any features that are recognized.

6.3.4 Metamorphic grade and age

Reaction zones develop only when two dissimilar rocks are adjacent to each other at certain specific conditions of metamorphism. Sometimes, reactive rocks are physically placed into contact by faulting or by intrusion and crystallization of magma at suitable metamorphic conditions. Sometimes potentially reactive rocks are deposited together as sediments, but no significant reaction occurs until they are brought to higher pressures and temperatures, at which point metamorphism occurs. Sometimes, minerals can exist in stable equilibrium at certain conditions of pressure and temperature, but react at others. For example, olivine-rich bands and plagioclase-pyroxene bands of a layered gabbro are stable together at high temperatures, such as those of igneous crystallisation, where they are part of an olivine+pyroxene+plagioclase assemblage. At low metamorphic grade, they metamorphose to two different rock types (serpentine and greenschist, respectively), which are not part of the same chemical equilibrium. At their contacts, they are liable to react to form a reaction zone of chlorite + tremolite. Metamorphic grade is vital in determining whether reaction between two rocks will occur, and what reaction zones will be produced.

Metamorphosed rocks do not always equilibrate internally to later conditions of metamorphism, particularly to those of a lower grade, where fluid addition is generally required to fully retrogress. However, reaction zones that experienced fluid infiltration at appropriate conditions can record the lower grade conditions. Grade discrepancies between reaction zones and neighbouring rocks may be useful in establishing a history of both metamorphism and displacements of rock masses. Where major thrusting or igneous activity has placed rocks together, the rocks on either side of the reaction zone may themselves differ in grade. It is even possible that a thermal gradient existed between the rocks on each side of the contact as reaction between them occurred. For all these reasons, the metamorphic grade of reaction-zones should be considered and it should be reported whether or not there is any obvious difference between their grade and that of their neighbouring rocks. If the zone is of lower grade, there may be other lower grade features (shear zones, perhaps) with which it may be correlated, or it may display a late fabric. Any case of a reaction zone of higher grade, or with grade changing from band to band through the zone, is unusual, and could be useful in geotectonic interpretations. Such cases should be highlighted, and described as fully as possible.

FAULTS, MYLONITES, AND CATACLASITES

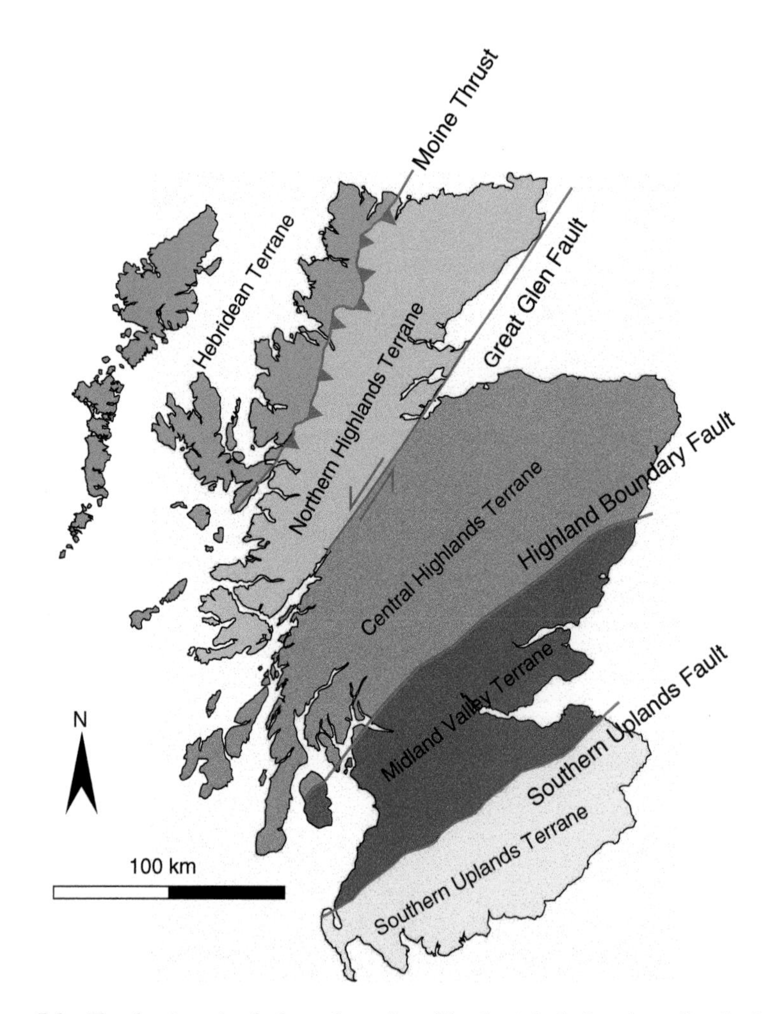

Figure 7.1 *The classic major faults and terrains of Scotland, including the strike-slip fault, the 'Great Glen Fault', the 'Highland Boundary Fault', and the 'Moine Thrust'.*

7
FAULTS, MYLONITES, AND CATACLASITES

In metamorphic terranes of all grades, rocks can be displaced and separated along discrete fault planes, intercalating faults/fault zones, and along more complex deformation/shear zones. It is clearly important when faults and fault/shear zones are found to record key structural data that helps to characterise their orientation, the direction of fault plane movement, and any shear sense indicators that may be used to work out dextral or sinistral shear within the system (see Section 7.3.2). Measurements should be unambiguous in their record, so that information can be easily reconstructed post fieldwork (e.g. correct compass declination). In many instances, it may be useful to take an orientated sample in the field so that thin sections or polished surfaces can later be made in the correct orientation. In order to do this you will need to mark on a dip and strike of a surface (and way up information), such that the subsequent thin section or polished slab can be orientated back in the lab. Faults and zones of shear are likely candidates for separating different metamorphic facies, and even major tectonic boundaries can be separated by relatively narrow zones of deformation, such that they map out as large-scale faults when considered at an appropriate scale (e.g. the Great Glen Fault, Moine Thrust, and Highland Boundary Fault, shown on Figure 7.1). When looking at fault rocks and faulted sections, it should be noted that the protolith concerned may have already been faulted many times before the main deformation phase that you are interested in, that faults are lines of weakness in a rock that can be utilised and reactivated again and again at different stages of the rock's history, and that deformation at different temperatures can impart strikingly different characteristics. Careful investigation of faults and shear zones can often identify preservation of different stages of deformation, therefore helping to piece together complex local or regional deformation histories. Where significant faulting and/or shear has occurred, specific fault rock types can develop, and can further help to classify and describe your examples.

7.1 Fault and Shear Zone Types

Displacement between two bodies of rock may be accomplished through any number of situations between the end-members of fault motion on a single plane and shear distributed across a wide zone separating two more coherent rock bodies. In general, the nature of rock deformation changes with depth in the crust as rocks go from very brittle (cold) conditions at shallow depths towards ductile (hot) conditions deeper, with possible partially melting at substantial depth or in the proximity of intrusions (e.g. Chapter 3, Figure 3.17). Due to this variation in conditions under which a rock may be deformed, there are a variety of different rock types associated with faults and shear. These are presented in Table 7.1, and examples will be presented and discussed in this chapter.

A conceptual way to look at the relationship of fault development with depth is depicted in Figure 7.2, which highlights the changes from brittle shallow crustal conditions, through the loosely defined 'brittle-ductile' transition and into the deeper ductile crust. The exact position of behaviour from brittle to ductile is not exactly the same for all rock types and situations, as different minerals have different temperatures and pressures under which they first behave in a ductile fashion. Some

The Field Description of Metamorphic Rocks, Second Edition. Dougal Jerram and Mark Caddick.

Table 7.1 *The classification of fault associated rock types.*

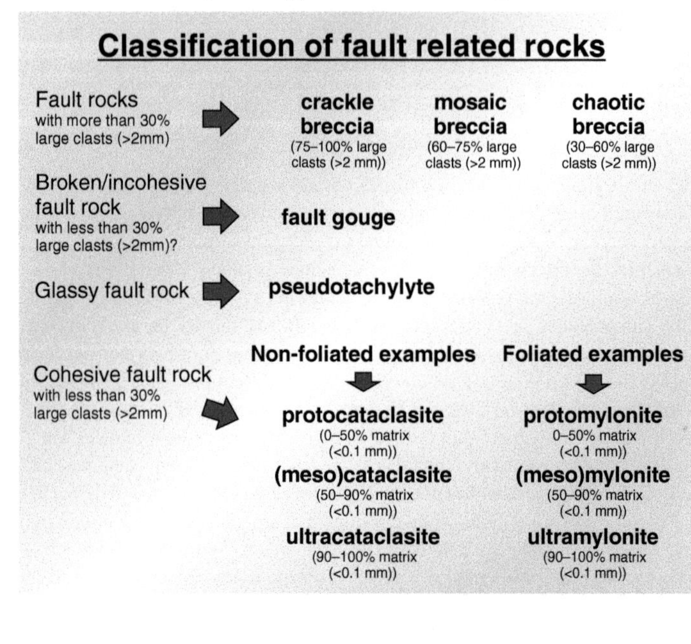

of the brittle-ductile transition temperatures for common minerals are also indicated on Figure 7.2. It should also be noted that the rates at which rocks and minerals are deformed also affect whether they respond in a brittle or ductile manner: think of the way that toffee flows if pushed slowly but shatters if struck with a hammer, and of the fact that it flows more readily (has lower viscosity) if warm.

7.2 Faults and Fault Breccias

In the simplest case and in the upper brittle parts of the crust, rocks can be deformed along discrete planes of weakness called faults. These are planar fractures within the rock mass that can accommodate movement between adjacent rock units without those units themselves having to experience substantial deformation. Pre-existing planar features such as joints or thin beds of 'weaker' minerals act as zones with a higher probability of developing into faults than the rocks around them. Any fault that can be observed in the field may have facilitated just centimetres of displacement or may have accommodated tens to hundreds of kilometres of movement during its history (e.g. the Great Glen fault, shown in Figure 7.1, likely experienced ~100 km of cumulative displacement). Marker beds, distinct lithologies, and broken mineral grains or fossils can all be used to estimate the extent of fault displacement. Note that sometimes a simple-looking fault may still have experienced significant movement/throw across it.

7.2.1 Slickensides and slickenfibres

When a fault surface is exposed parallel to the outcrop surface, it often consists of a smoothed surface with lines and ridges (striations) along it (Figure 7.3). These features, known as slickensides, can help to depict the movement direction of the fault. The striations themselves are typically oriented parallel to the direction of fault displacement, with steps in the polished surface revealing the

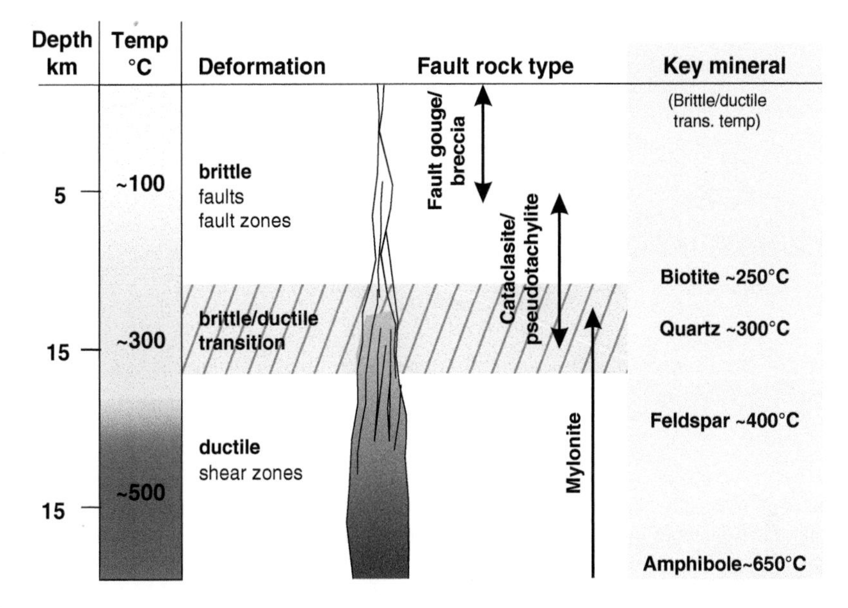

Figure 7.2 Schematic to highlight changes from ductile to brittle behaviour in rocks and the approximate temperatures of the brittle-ductile transition in important minerals (in part adapted from Nike Norton [CC BY-SA 3.0]).

Figure 7.3 Quartz slickenfibres with congruous steps in quartzite, indicating hanging-wall motion towards the reader. Arkaroola Creek, South Australia (photo courtesy of Christoph Schrank).

polarity of this movement: the surface feels smooth when a finger or hand is brushed along it in the same direction that the now eroded side of the fault moved, and rough when brushed in the opposite direction (e.g. Figure 7.4).

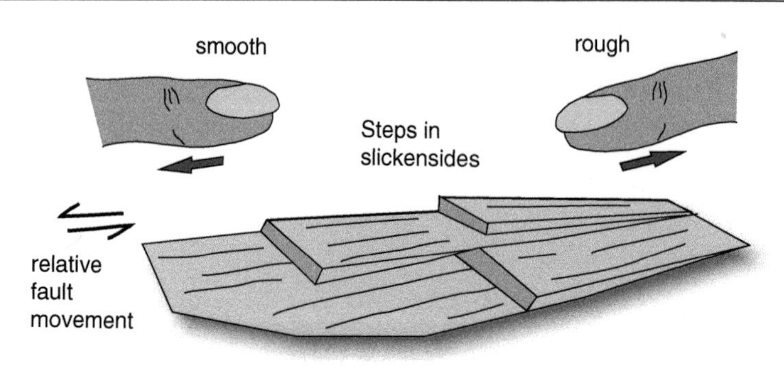

Figure 7.4 *Schematic to show how steps on slickenside surface can be interpreted to reveal displacement direction (see also Figure 7.3).*

Faults often accumulate or focus fluids, with mineral precipitation common in fault zones. Growth of new minerals during active faulting typically results in alignment of new crystals with the direction of fault motion, with these highly aligned crystal bundles termed slickenfibres and enhancing the rough/smooth nature of the fault surface (as described in Figure 7.4). Both slickensides and fibres can be very valuable in determining relative fault motion and type (reverse, normal, oblique, etc.), and are particularly useful in cases where movement has been sufficient that comparison of both sides of the fault does not reveal obviously displaced marker beds. Because slickensides and fibres form lineations, their plunge and a plunge direction (azimuth) should be measured and recorded, as should the dip-strike and dip direction of the fault plane. If smooth and rough directions along the surface can be determined, then the type of throw on the fault can also be recorded.

Faulting is common at very shallow depths in tectonically active regions, so the displacements recorded by a brittle fault may have formed much later than the higher temperature and pressure conditions at which metamorphic minerals and fabrics were produced in the wall rocks. The sense of shear of these later faults may even be entirely different to earlier displacement at greater depth. To aid with unravelling whether this is the case, the mineral species (where discernible fibres are present) and sense of movement recorded by slickensides/fibres should always be recorded. As many examples as possible should be noted in order to help determine whether any single brittle fault is of broad tectonic significance (be careful – slickensides/fibres may reveal only that large rock masses are actively creeping down present-day hillsides along minor faults).

7.2.2 Chemical changes at and near faults

Faults can accommodate substantial deformation, can focus fluid flow, and can juxtapose unlike rocks. Chemical processes are therefore likely to affect the fault rocks and the wall rocks to some distance each side of a fault. In the field, such processes may be very obvious due to mineral colouration within the fault zone (e.g. Figure 7.5). Around the fault, look out for cloudy alterations of feldspars, reddening from oxidation and precipitation of ferric oxides and hydroxides, carbonate replacement of igneous rocks, and dull chloritic alterations to schists. Even where chemical effects are not noticeable, there may be changes in weathering, or in the character of joints, which should be recorded.

Figure 7.5 *Epidote formation along an E–W cataclastic fault zone within a Palaeoproterozoic orthogneiss, Praia das Conchas, Cabo Frio, RJ, SE Brazil (photo courtesy of Isabel Carmo).*

7.2.3 Fault breccia – gouge

Major brittle faults often contain areas of non-coherent 'crumbly' breccias, along with brecciated sections that have been cemented with secondary minerals. These fault breccias are the result of continued fragmentation of the rock mass into angular chips during successive movements along the fault, and can connect more cohesive segments of fault planes. The breccia will often contain large fragments in a finer matrix, and many secondary minerals such as calcite, quartz, and gypsum can precipitate in gaps. The term 'fault gouge' can be used to describe a fault breccia in which a significant reduction in the grain size of the breccia has occurred, often forming a muddier groundmass by significant grinding and milling of the faulted material. Fault gouge can also be used to depict a degree of non-cohesiveness (e.g. Table 7.1), though some degree of secondary cementation can occur. Previously, fault breccia was considered a non-cohesive unit, but this caused some difficulty in the field due to cemented and coherent breccia's, so now the degree of clasts and clast size is used in the definition (Table 7.1). Some examples of fault breccias are given in Figure 7.6.

7.3 Cataclasites and Pseudotachylites

At slightly deeper levels of the crust and with continued development of faults under greater temperature and pressure, fault breccias tend to further reduce in grain size and are also generally more coherent, leading to a group of fault rocks known as cataclasites. Rapid faulting under pressure can lead to the formation of pseudotachylites, sometimes interpreted to reflect rapid and localised melting.

7.3.1 Cataclasites

These fault rocks develop when a significant proportion of the breccia matrix has been reduced in grain size during multiple fracturing events as the fault system has developed. This continued fracturing or 'cataclasis' leads to a point at which further deformation along the fault can be accommodated by the rolling or sliding of grains past each other, in a sort of flow of the breccia matrix

5. Understanding Textures and Fabrics 2
6. Contacts, Reaction Zones, and Veins
7. Faults, Mylonites, and Cataclasites
8. Summary Tables, Checklists, and Mapping Report Advice

Figure 7.6 *(a) Crackle to mosaic breccia of Silurian mudstone with calcite cement. Dent Fault Zone, River Rawthey, Cumbria. (b) Crackle to mosaic breccia of Devonian limestone clasts in calcite cement. Hope's Nose, Torquay, Devon. Breccation is along successive weakly aligned extension vein systems (photos courtesy of Nigel Woodcock).*

material. The type of cataclasite depends on the percentage of matrix material (see Table 7.1), but generally there is no fabric developed within the fault rock. Protocataclasite contains < 50% matrix, (meso)cataclasite contains 50–90% matrix and ultracataclasite describes a faulted rock reduced to > 90% fine matrix (see Table 7.1). Figure 7.7 highlights some examples of cataclasites in the field (see also Table 7.1 and Figure 7.2 for classification and relative depth of occurrence, and the example shown in Figure 7.5).

Figure 7.7 *Cataclasite examples. (a) Brecciated greywacke with thin sub-horizontal layer of cataclasite (ca. 5 cm underneath coin), Axial Ranges, North Island, New Zealand. (b) Layer of cataclasite (marked by coin) embedded in brecciated schist, Paralana fault, Arkaroola Creek, South Australia (photos courtesy of Christoph Schrank).*

7.3.2 Pseudotachylites

These are enigmatic fault rocks, often interpreted to represent a 'frozen' melt along the fault plane, in veins and in-filling (back veining) fault breccia frameworks. The melt is rapidly generated and quenched to a glass. Pseudotachylites are often black like obsidian or 'tachylite' glass, though many examples are partially or wholly devitrified into a very fine-grained dark rock. They often contain fragments of the surrounding wall rocks within them, and may show evidence of quench cooling.

Figure 7.8 *Thin layer of pseudotachylite with apophyses (glassy black material below coin) in polydeformed schist, Cap De Creus, Spain (photo courtesy of Christoph Schrank).*

Pseudotachylites are generally considered to be formed by frictional melting during rapid motion, associated for instance with earthquakes, caldera collapse faults, meteorite impact structures, and occasionally along some major landslides. Most commonly they are attributed to seismic faulting episodes. Examples of pseudotachylite textures are given in Figure 7.8.

7.4 Mylonites and Shear Zones

Deeper in the crust, where rock masses can deform in a ductile fashion, wider shear zones develop and can lead to a type of fault rock known as mylonite (see Figure 7.2). These rocks can be tricky to deal with in the field, but are an extremely important sub-group of fault rocks and can have significant implications for understanding the deformation history and major structural boundaries of your metamorphic sequence.

7.4.1 Mylonite rocks

Mylonites are medium- to fine-grained metamorphic rocks that generally develop in ductile shear zones where significant strain is focused. They form mainly during simple shear through dynamic recrystallisation and crystal-plastic deformation of grains within a rock. A mylonitic schist or gneiss has had its grain size reduced towards the fineness required for continuing ductility at the physical conditions of the particular deformation. A banded rock within a coarser host should be termed 'mylonitic' if it can be shown to occur in a definable deformation zone, or if it contains porphyroclasts and/or other key indicators of shear (see also Chapter 5).

Particularly significant identification problems are sometimes associated with fine-grained rocks of the mylonite series that have felsic or intermediate plutonic or gneissic protoliths. Some are slates or phyllites, while others are chert-like. Those with scattered and rounded-off broken grains of quartz and feldspar can appear meta-volcanic or volcaniclastic. With finer-grained examples, the identification of mineral lineation, C-S fabrics on the correct surfaces (see shear sense indicators in Section 7.4.2) and small rotated grains can help determine the orientation of shear.

However, in many cases, detailed thin section work might be the best way to reveal the details of their mylonitic nature from carefully sampled, orientated specimens. Granitic rocks with only protomylonitic alteration of the edges and corners of grains can be mistaken for arkosic metasediments. If these contain chunks of rock which have escaped alteration, they may appear to be metaconglomerates. When encountering mylonites in the field it is important to assess the shear direction to help determine how they fit into the broader tectonic and metamorphic history of the field area.

7.4.2 Shear sense indicators in mylonites

As their name indicates, shear zones are regions of rock that have undergone shearing (dextral, sinistral, thrusting, normal, or composite) through focusing of strain. These zones lead to the development of mylonitic rocks and can preserve key metamorphic textures that indicate the shearing motion that they have been deformed by, and the timing of this deformation relative to metamorphic mineral growth. We have already touched on some of the key textural observations, both planar and as isolated bodies within the metamorphic rock (e.g. Chapters 4 and 5), that are of use in determining and quantifying structures within a shear zone. As many of the textural elements are of particular importance within mylonites, it is useful to revisit these and also to consider the correct way in which that data should be recorded in the field (Figure 7.9).

Textures that are produced in simple shear have an asymmetry which allows you to determine the shear direction, but asymmetric structures can also be formed in several ways, so great care is needed to record the data correctly. First and foremost you will need to be looking at the correct surfaces to see textures that are representative of the shear motion within a deformed zone: try to primarily make observations from rock surfaces that are parallel with the direction of shearing and perpendicular to the foliation (as indicated in Figure 7.9). You will need to spend some time getting to grips with your outcrop, determining the best exposure faces, becoming familiar with lithological

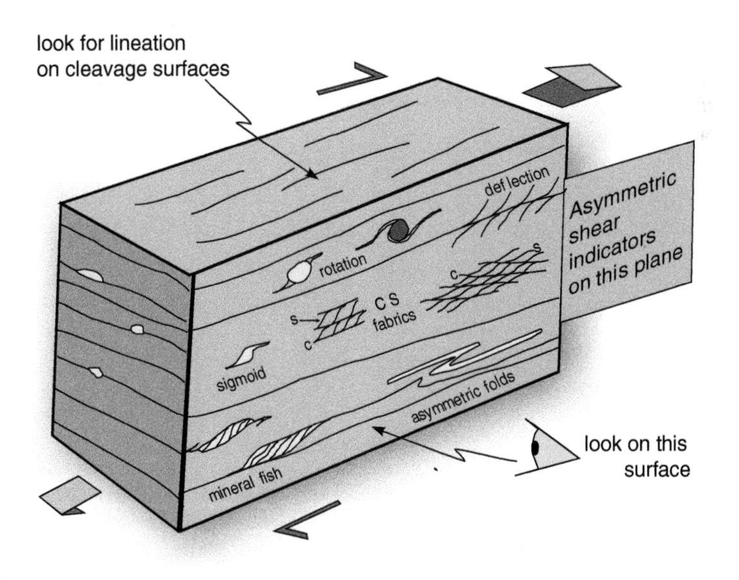

Figure 7.9 *Diagrammatic representation of the plane of orientation to look for shear sense indicators in sheared rocks.*

variation and identifying as many shear-sense indicators as possible. Some lithologies better preserve certain asymmetric structures, so you may need to identify several key sites to measure different types of shear sense indicators in order to build a good quantitative picture of shear motion that best characterises the shear zone (see examples in Figure 7.10). Many of the indicators of shear and fault/deformation movement are often found within rock textures, and annotated examples and 'mini' sketches are worth recording in your notebook (e.g. Figure 7.11). *NOTE – shear zones are*

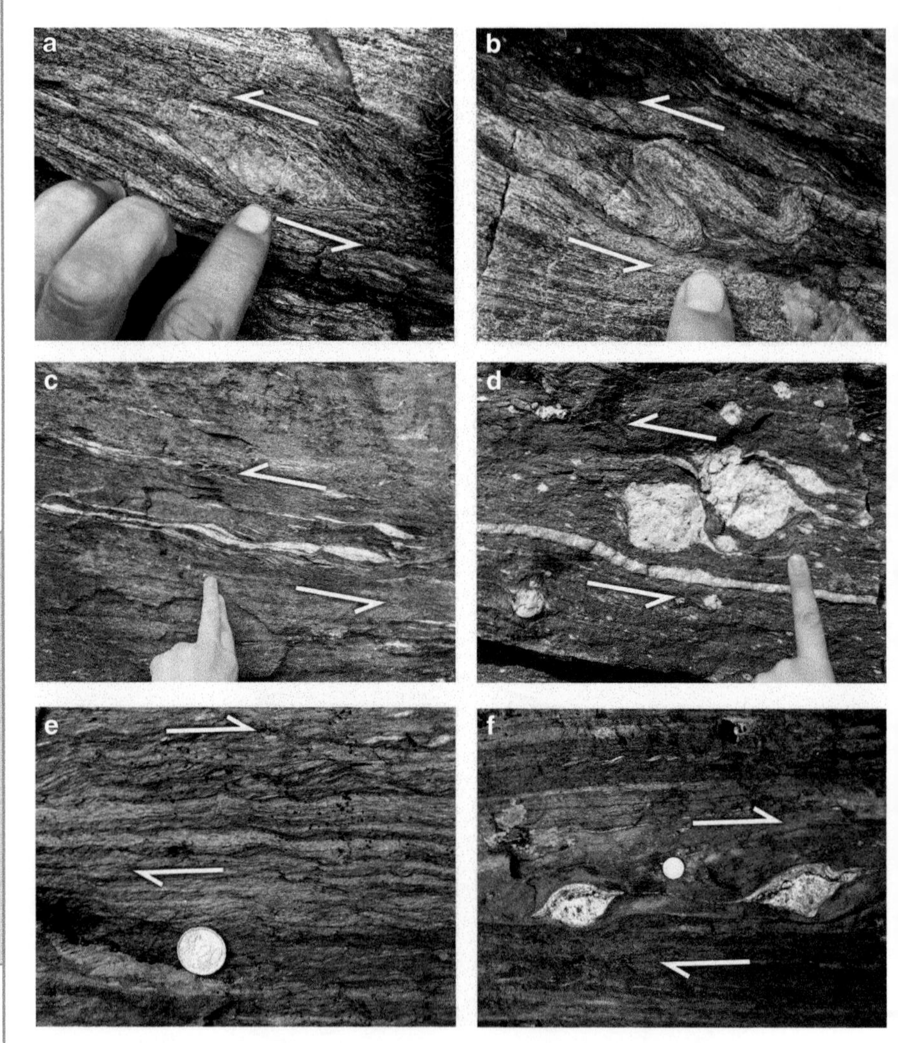

Figure 7.10 *Examples of shear indicators. (a) Sinistral shear in Moint Trust shear zone Assynt, Scotland. (b) Fold in Moine Thrust shear zone (sinistral), Assynt, Scotland. (c) Large porphyroclasts of feldspar entangled in the Norfjord Sogn detachment mylonites, Norway. (d) Shear bands deforming a light coloured band in an orthogneiss, Norway. (e) Dextral SC' shear bands in schist, Cap de Creus, Spain. (f) Dextral sigma clasts in mylonitic shear zone, Cap de Creus, Spain (photos: (a) and (b) Dougal Jerram; (c) and (d) courtesy of Hans Jørgen; (e) and (f) courtesy of Christoph Schrank).*

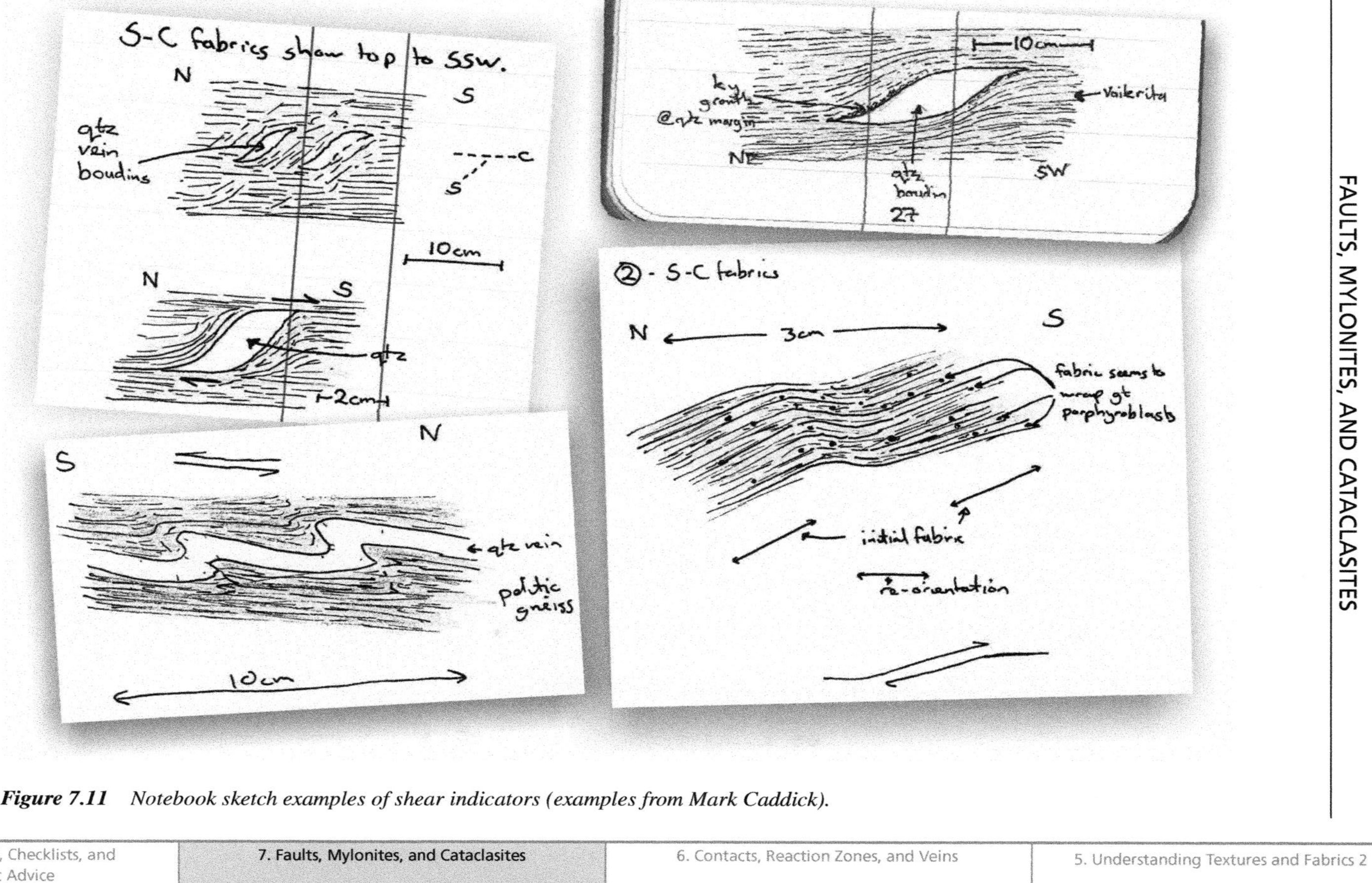

Figure 7.11 *Notebook sketch examples of shear indicators (examples from Mark Caddick).*

areas of weakness in the crust that can be reused again and again (so-called fault reactivation), so ideally you should collect as much structural data as possible to best characterise the main shear event, bearing in mind that later events may have a different shear direction. It is not uncommon to find contrasting shear sense indicators within any given shear zone and to eventually discover evidence that textures associated with one of the sets are overprinted by a second set, which therefore represents a later phase of deformation with a different orientation.

Figure 7.12 *Examples of sheath folds, Assynt, Scotland. Umbrella in (b) and (c) used to indicate 3D structure that is sectioned in the 2D rock face (photos Dougal Jerram).*

Folding is common in mylonites, mylonitic schists, and mylonitic gneisses, and sometimes takes a sheath-fold form (a complex three-dimensional shape resembling a cone or sheath/condom). The folds are produced by several shearing episodes in which early-formed folds are themselves folded again, a little like pushing folds along a table cloth to produce complex fold morphologies. In cross-section along certain orientations you can see a circular or elliptical closure of bands where the sheath form is truncated (e.g. Figure 7.12). Fold shapes should be sketched, particularly on a number of differently oriented exposure surfaces, and the three-dimensional fold shapes and relationships measured (as in Chapter 4).

Ultimately you are likely to be faced with a number of planar fractures within a rock that may be simple joint features, fairly localised faults, or more extensive faults. Additionally, 'fault zones' may be recognised where the faulting is somewhat more important or clearly manifest in the rock mass. The identification of 'shear zones' will also help highlight areas where significant rock deformation has occurred. In each case, you will need to try to assess the relative importance of the structures and their relevance to the overall metamorphism and deformation within the area. It should be noted that while most major faults will be obvious, even some major faults such as thrusts can be located on fairly discrete planes, and can be somewhat surprisingly inconspicuous.

SUMMARY TABLES, CHECKLISTS, AND MAPPING REPORT ADVICE

Table 8.1 *An overview of the key rock types formed from the major protolith groups upon progressive metamorphism. Colour categories for each protolith are used elsewhere in this book, as are the grade codes. These codes are very low grade (VLG), low grade (LG), medium grade at low to intermediate pressure (MG1), high grade at low to intermediate pressure (HG1), very high grade (VHG), medium grade at high pressure (MG2) and high grade at high pressure (HG2).*

	Grade code	General Classification of Protolith Type: A) Hydrated Ultramafic	B) Mafic igneous	C) Felsic igneous	D) Psammitic Sediments	E) Pelitic Sediments	F) Carbonates	G) Semi-carbonate	Grade code
Unmetamorphosed		Serpentinite	Basalt	Rhyolite/ Granite	Sandstone	Shale	Limestone/ dolomite	Marl or mixed protolith	
Increasing metamorphic temperature at crustal and thickened crustal depths	VLG		Zeolite	Little change without fluid addition or deformation →	Increasing grainsize Increasing bulk hardness →	Slate	Increasing grainsize Increasing bulk hardness →	Calcareous Slate	VLG
	LG	Talc schist	Prehnite/ Pumpellyite			Phyllite		Increasing grainsize, isotropy & bulk →	LG
	MG1	Amphibole Peridotite	Greenschist			Schist			MG1
	HG1		Amphibolite			Gneiss			HG1
	VHG	Peridotite	Granulite		Quartzite	Migmatite	Marble	Diopside-gneisses	VHG
Increasing temperature at high pressure		Serpentinite	Basalt	Rhyolite/ Granite	Sandstone	Shale	Limestone/ dolomite	Calcareous Slate	
	MG2	Chlorite-Peridotite	Blueschist			Micaceous Blueschist		Calcareous Blueschist	MG2
	HG2	Peridotite	Eclogite		Quartzite	Micaceous Eclogite Gneiss	Marble	Calcareous Eclogite	HG2

8
SUMMARY TABLES, CHECKLISTS, AND MAPPING REPORT ADVICE

The previous chapters have given you the information and hopefully imparted some of the skills and knowledge that you will need when working within metamorphic terrains, and areas associated with metamorphic rocks in combination with others (sedimentary/igneous). As you improve and hone your field skills, you may want to use a quick reference to remind you of the most relevant and important minerals, rock types, and associations for different protoliths and metamorphic grades. This chapter aims to act as a summary reference and checklist, providing tables of mineral and rock names and properties to aid your work. We use summaries of protolith groups (compositional categories) along with metamorphic grade indicators to help you navigate through the main rock types and their constituent minerals. Additionally, more advice about mapping and final report structures, as well as more information on geologic formations and markers, are given at the end.

8.1 Compositional Categories and Their Grade Indicators

Here we outline the compositions that represent the majority of metamorphic rocks you are likely to be interested in. For each, we discuss some of their diagnostic features, their likely origin, and some key mineral associations at various grades. The compositional groups described below generally refer to an unmetamorphosed protolith type, using the same letter codes for each group that were adopted in Chapter 3. We emphasise that protolith compositions can be substantially modified by metasomatism, melt loss, or melt addition, though often only in a localized and interpretable sense (see for example, Sections 6.1.4 and 6.3). Metamorphic grades discussed here are grouped as in Chapter 3, which you can use as a cross reference. We have also repeated the key protolith and grade table used in Chapter 3, here as Table 8.1 on the inset page, to ease cross referencing within this summary chapter. The grades used are:

- Very Low Grade – *VLG*
- Low Grade– *LG*
- Medium Grade – *MG*　　At subduction pressures – *MG2*
- High Grade – just *HG*　　At subduction pressures – *HG2*
- High Grade – *HG*
- Very High Grade – *VHG*

A. *Ultramafic*

Diagnostics: Rocks rich in some of: *serpentine, talc, olivine, anthophyllite, cummingtonite, enstatite, bronzite*. This includes *SERPENTINES* and *PERIDOTITES.*

Derivation: Ultramafic igneous rocks. Very rarely from associated very local and very immature sediments from an ultramafic source. *Often serpentinised before metamorphism.*

The Field Description of Metamorphic Rocks, Second Edition. Dougal Jerram and Mark Caddick.

Mineral origins: Bronzite, augite, and *chromite* are possible igneous remnants, while *olivine, plagioclase, spinel, garnet*, and *magnetite* may be either igneous or metamorphic. Remnant igneous minerals need not accord with metamorphic grade. In addition to the diagnostic minerals and those mentioned above, the following metamorphic minerals may occur: *magnesite, brucite, chlorite, diopside, tremolite, phlogopite*.

Grade: These are good rocks for indicating grade category. Serpentinites are only stable at *VLG* or *LG*. Peridotites are *MG*, *HG*, *VHG*, or igneous. Abundant *talc* is usually *MG*. Abundant *anthophyllite, cummingtonite*, or *enstatite* reflects *HG*.

For more precise estimates of grade, search for talc or magnesite bearing rocks and for siliceous contacts, searching for these mineral combinations:

Serpentine + quartz (without talc) – *VLG*

Serpentine + talc (any *quartz* is *talc-coated*) – *LG*

Olivine + talc – *MG*

(Anthophyllite or *cummingtonite) + (olivine* or *talc)* – just *HG*

(Anthophyllite or *cummingtonite) + (enstatite* or *quartz)* – *HG*

Periclase; Spinel; Sapphirine – *VHG*

Subdivision of *LG* and *MG* is possible by these combinations:

Serpentine + brucite – *VLG* or *lower LG*

Serpentine + olivine – *upper LG*

Olivine + talc + magnesite – *lower MG*

Olivine + talc + scattered anthophyllite or enstatite – *upper MG*

Chlorite, phlogopite, tremolite, and *diopside* are not particularly good grade indicators in ultramafic rocks.

B. *Mafic (also sometimes called basic) and intermediate*

Diagnostics: Rocks having *mafic* and *calc-aluminous* portions, and sometimes also containing *quartz* and *biotite*. This includes *GREENSCHISTS* and *AMPHIBOLITES*.

The *mafic portion* typically consists of: *chlorite, amphiboles, pyroxenes, garnet*.

The *calc-aluminous* typically consists of: *plagioclase, albite, epidote group, zeolites, prehnite, pumpellyite, lawsonite*.

Derivation: Mafic (basic) or intermediate igneous rocks. Sometimes immature sediments sourced from these rock types. Widespread *carbonate* may occur if sediments or lavas were subjected to hydrothermal alteration. Extremely rarely (if *MG* or higher) from dolostones and marls (see **G** and Section 3.5.6). See Section 3.5.3.

Grade: These rocks give an indication of grade category, as shown in Tables 8.1 and 8.2 and associations comprising the classic facies scheme (e.g. Figure 1.2). However, changes at category boundaries may be gradational.

Biotite can occur at *LG*, *MG*, and *HG*, while *chlorite* is typically not present in rocks of *HG* or *VHG*. *Actinolitic* amphiboles give way to *hornblende* at *HG*.

Garnet or *diopside* may be usable for defining zones in the field (using either their first appearances or mineral proportions) but are not a universally reliable grade indicator. *Cummingtonite* may occur in hornblende-rich rocks. Relict igneous minerals may remain and need not accord with metamorphic grade.

VLG: A colour sequence of *white* (zeolites, analcite) to *green* (prehnite, pumpellyite) to *yellow-green* (epidote), in fine-grained igneous materials, is prograde within *VLG*, but may result from heating in a lava pile, not from later metamorphism.

C. *Felsic (also sometimes called acid or silicic)*

Diagnostics: Much *quartz* and *K-feldspar*, and sometimes *albite* (or *plagioclase*). Sometimes *mica*-bearing. Can include some *epidote, chlorite, amphibole, zeolite, prehnite*, or *stilpnomelane*.

If too quartz-rich for an igneous rocks: see **D**.

If too micaceous for an igneous rock: see **E**.

Derivation: Generally from felsic igneous rocks though sometimes from granitic conglomerates or arkosic sands and granule-conglomerates. Distinguishing between these is generally possible because coarse igneous or sedimentary textures are normally retained unless deformation has been intense.

Grade: These rock types are almost useless for indicating grade. However, certain minerals, if present, give indications of grade:

Chlorite; epidote	– *VLG* or *LG*
Stilpnomelane	– *VLG* / *LG* boundary
Biotite	– *LG*, *MG*, or *HG*
Hypersthene with garnet, cordierite, sillimanite; or Sapphirine	– *VHG*

If migmatised, these rocks reflect HG or VHG.

D. *Quartzite*

Diagnostics: Dominated by quartz. This includes *QUARTZITES*, including *ARKOSIC* and *MICACEOUS* varieties.

Derivation: Protoliths are mature sorted sandstones and conglomerates, and the name 'psammite' (the Greek-derived equivalent to the term arenite) is commonly used when metamorphosed.

Grade: Generally useless for indicating grade, except if containing:

White mica	– *VLG*, *LG*, or *MG*
Al-silicate	– *HG* or *VHG*

(See also category **E**. Note presence or absence of K-feldspar.)

E. *Semi-pelite and pelite*

Diagnostics: Rich in *sericite* or *mica*, or containing an *aluminous mineral*. Other common minerals include *quartz, chlorite, feldspars, garnet*.

The aluminous minerals are typically chloritoid, staurolite, cordierite, and *Al-silicates* (*andalusite, kyanite, sillimanite*).

This group includes most *MICASCHISTS* and *SLATES*, also some gneisses.

Derivation: A wide range of clastic sediments having a high proportion of fine-grained material. Broadly speaking: greywackes, siltstones, and mudstones. ('Pelite' = argillite.)

Grade: Although these rocks show a number of changes with increasing grade, category boundaries can be ill-defined unless they are either pelites proper (described separately below as **E+**) or *HG* has been attained. Note that *biotite* normally appears at some point within *lower* LG, and that *quartz* may exist at all grades. Any *stilpnomelane* indicates VLG/LG *boundary*. At VHG, *K-feldspar* (and *plagioclase* and sometimes *quartz*) *may be lacking*, indicating loss of alkali metals into partial melts. At LG and MG, *K-feldspar* can occur only in the absence of aluminous minerals (i.e. in semi-pelites not in pelites).

Some minerals in these rocks are PRESSURE INDICATORS (see also Figures 3.9 and 3.10):

Garnet	– intermediate or high *P*
Cordierite	– intermediate or low *P*
Kyanite	– high *P*. (Extreme high *P* if HG)
Andalusite	– intermediate or low *P*. (Low *P* if HG)

Further subdivision is possible for pelites proper, as described in E+.

E+. *Pelite subgroup*

Diagnostics: Chloritoid; Staurolite; or the combination: *Aluminous mineral + white-mica + quartz*.
Derivation: Chemically mature mudstones.
Grade: Some grade indicators for group E+ can be derived from the AFM triangles in Figure 3.10. In addition to the indicators given under **E**:
Indicators of LG: The presence of *chloritoid or garnet + chlorite;* restriction of any *kyanite* or *andalusite* to biotite-free rocks. Within LG, *garnet* indicates *upper* LG.
Indicators of MG: *Staurolite; Cordierite (+ white-mica + quartz); Al-silicate + biotite (+ white mica + quartz)*. Within MG:

Staurolite + chlorite	– *lower* MG
Al-silicate + biotite	– *upper* MG*; or low P; or both*
Sillimanite; Garnet + cordierite	– *upper* MG *at* intermediate *P*
Staurolite	– Intermediate or high *P*; or low *P* at lowest MG

Migmatisation, the loss of muscovite and formation of K-feldspar, or the formation of orthopyroxene all imply HG.

F. *Calcareous*

Diagnostics: Rocks rich in carbonate minerals. *MARBLES*.
Derivation: Limestones (this does not deal with hydrothermal or metasomatic deposits, with magnesite in association with evaporites or serpentines, or with sideritic sedimentary or metasedimentary rocks).
Minerals in marbles: Pure calcite limestones metamorphose to marbles but do not change with metamorphic grade. Non-carbonate minerals in marbles usually represent detrital or evaporitic impurities *(quartz, chlorite, sericite, mica, gypsum)*, which in larger quantities generate calc-silicates (category **G**). Dolomites with low original clay content but a substantial amount of silica (normally chert) metamorphose to Mg-silicate-bearing calcite marbles.
Grade: Marbles containing calc-aluminous or magnesian silicates can be good grade indicators, but their treatment is not easy in the field. Metamorphosed siliceous dolomites in particular are of somewhat restricted occurrences, and their minerals are not always readily identifiable.
Shaly or slaty fine-grained marbles indicate VLG.

Chloritic or micaceous marbles indicate VLG or LG.
Calc-aluminous minerals in marbles – as in **G**.
The grade indicators given below occur in metamorphosed siliceous dolomites (now calcite-rich marbles); see Section 3.5.5. The sequence of first appearances: *talc, tremolite, diopside, olivine* spans LG and MG, but it is possible for *talc* and sometimes also *tremolite* to be bypassed. *Chlorite* is not a reliable grade indicator in such rocks. The following minerals indicate the grade shown, *if coexisting with calcite* (but their absence does not refute such a grade):

Talc; Phlogopite; Dolomite + quartz	– VLG *or* LG
Tremolite; Actinolite	– MG
Wollastonite; Periclase; Spinel	– HG
Olivine; Dolomite + diopside; Scapolite	– MG *or* HG

At shallow igneous contacts, a series of zones of unusual minerals may be discernible. The first of these minerals are often *monticellite* and *melilite*, but identification is not usually possible in the field.

G. *Meta-marls and calc-silicates*

Diagnostics: Either rich in both *calcite* and *silicates*, or rich in Ca-silicates such as *diopside, plagioclase, grossular, epidote,* or *wollastonite*. See Sections 3.5.5–3.5.7.
Derivation: Sedimentary mixtures of carbonate and clastic materials.
Grade: Calc-silicates are theoretically good grade indicators, but in practice there may be difficulties with the mineral identifications needed. The major reactions in them do not coincide precisely with the grade category boundaries previously defined.
VLG: a mixture of carbonates and the silicates of VLG metaclastic rocks (*chlorite, sericite, quartz*), normally as a calcareous slate.
LG: a mixture of carbonates and the silicates found in various rocks at this grade (*white mica, chlorite, quartz, albite, epidote, actinolite, biotite*).
Upper LG, MG, and HG: very variable assemblages, commonly containing *diopside*. *Biotite, actinolite, diopside*, and *garnet* can each persist throughout these grades (particularly in patches where all *calcite* has been used up). Diopside-bearing amphibolites may occur (Section 3.5.6). The following grades are indicated by *coexistence with calcite:*

Chloritoid	– *lower* LG
Zoisite; Clinozoisite	– LG *or lower* MG
Plagioclase + white mica	– *upper* LG *or lower* MG
K-feldspar + plagioclase + quartz	– *upper* MG
Grossular + anorthite + quartz	– MG at low P; HG at high P
Wollastonite; Spinel	– HG

H. *Rocks of high pressure of all compositional types*

Diagnostics: Blue amphiboles; Green (jadeitic or omphacitic) pyroxenes; Greenish or *brownish phengite micas; Lawsonite; Aragonite*. Blue amphiboles combine NaAl with a mafic constituent. High pressure pyroxenes contain NaAl in solid solution with either CaMg of diopside, or NaFe of aegirine, or both. Phengites are mafic rather than aluminous, having compositions similar to (K-feldspar + biotite), (K-feldspar + chlorite) or (K-feldspar + orthopyroxene). Therefore, the normal compositional categories of rocks (**A – G**) are less suitable for their subdivision.

Ha *Marginal blueschist association*

Blueschists or *blue amphibole-bearing greenschists* (usually fine-grained and of mafic igneous origin) *occur in* VLG, LG, *and* MG2 *rocks*. A little *blue amphibole* may occur in metamorphosed plutonic rocks, marbles, and metaclastic rocks.

Blue amphibole may exist with *albite, chlorite, phengite, stilpnomelane, calcite, quartz, sphene, actinolite, epidote, lawsonite*. If CaAl-silicates can be identified, they indicate:

Lawsonite – marginal to VLG to lower MG2

Zoisite; Clinozoisite – marginal to LG

Hb *Jadeite blueschist association*

Blue amphiboles are *widespread*, with concentrations in metabasic rocks. *Jadeitic pyroxene* occurs with *quartz* in felsic and some mafic igneous rocks, and in some metasedimentary rocks. *Albite* is typically unstable (having reacted to form a jadeitic pyroxene component). *Phengite* is common in felsic and pelitic rocks. *Lawsonite* in some basic and intermediate rocks and some marbles. *Aragonite* may occur, but may have reverted to *calcite*.

Other minerals: *stilpnomelane, garnet, sphene, actinolite, quartz*.

Indicates VLG to lower MG2 (high pressure).

Hc *Glaucophane eclogite association*

Blue amphibole, omphacite, garnet, phengite, white mica (Na-mica), *carbonates, quartz*, and *rutile* can all occur together in rocks of many compositions (including mafic and metasedimentary types). Mafic rocks also contain *zoisite, clinozoisite*, sometimes *green epidote*, and sometimes *green amphibole*. *Phengite, jadeitic pyroxene, quartz*, and *garnet* may occur in felsic and sedimentary compositions. *Blue amphibole, chloritoid*, and *epidote* in some schists. Other combinations of the minerals cited.

Indicates MG2 *to* HG2.

Hd *Eclogite association*

Mafic rocks of essentially *omphacite + garnet ('Eclogite')*, plus other minerals in small to moderate amounts, plus *rutile*. Other minerals can indicate variations of basic composition, e.g. *clinozoisite* where originally plagioclase-rich. The following accompanying mineral associations may indicate grade of metamorphism:

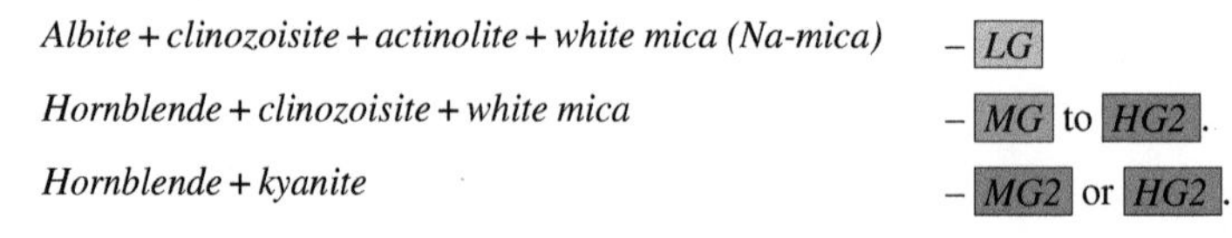

Albite + clinozoisite + actinolite + white mica (Na-mica) – LG

Hornblende + clinozoisite + white mica – MG to HG2.

Hornblende + kyanite – MG2 or HG2.

Eclogites have equilibrated at high pressure, at the grade indicated by additional minerals.

Warning: Some eclogites contain *olivine* plus minerals of HG or VHG, and are probably mantle rocks, not produced by metamorphism of crustal material. In others, low pressure minerals occur, generally as a result of late-stage hydrous alteration during exhumation, and are out of equilibrium with the eclogite.

8.2 Minerals

A list of selected metamorphic minerals and their usual properties.

Table 8.2 *Main mineral constituents of some compositional categories at different grades.*

Grade Category	MATERIAL			
	A Ultramafic	***B Mafic igneous:*** ***Mafic portion***	***Calc-aluminous portion***	***E Semi-pelite and pelite***
VLG	Serpentine (Quartz, magnesite)	Clays, chlorite (igneous relics)	Zeolite, pumpellyite, epidote, albite	Clays, chlorite, sericite, quartz
LG	Serpentine (Talc, magnesite)	Chlorite, actinolite (Garnet)	Epidote, albite	White mica, chlorite, quartz, biotite (Garnet, Al-minerals)
MG	Olivine, talc (Magnesite, anthophyllite)	Hornblende (Diopside, garnet)	Plagioclase	White mica, biotite, quartz (Garnet, Al-minerals)
HG	Olivine, anthophyllite, cummingtonite, enstatite	Hornblende (Diopside, garnet)	Plagioclase	K-feldspar, biotite, quartz, Al-minerals, (Garnet) —or migmatites—
VHG	Olivine, enstatite	Hypersthene, diopside (Hornblende)	Plagioclase	Hypersthene+Al-minerals (K-feldspar, quartz) or Sapphirine+other minerals

5. Understanding Textures and Fabrics 2
6. Contacts, Reaction Zones, and Veins
7. Faults, Mylonites, and Cataclasites
8. Summary Tables, Checklists, and Mapping Report Advice

8.2.1 Main mineral groups and other common bulk minerals (Table 8.3)

Table 8.3 *Main mineral groups and other common bulk minerals.*

Mineral (hardness)	Description	Compositional category (see 8.1)
Quartz (7)	Glassy. Colourless, except for a purple or blue—grey hue at VHG. Not subject to alteration. Trigonal.	Bold&highlight implies common mineral **All**
Carbonates		
Rhombohedral:	(Rhombohedral cleavages) Colour / Acid reaction / Weathering	
Calcite (3)	white / effervesces / weathers clean (rusty if ferroan)	**BFG**
Siderite (4)	brown / reacts quietly / rusty stain	
Magnesite (4) Dolomite (4) Ankerite (4)	white to yellow to brown	**AF**G
Orthorhombic:	(one cleavage)	
Aragonite (4)	White / effervesces / grey/white or cream	H
Feldspars (6)	Grey, white or cream. Equidimensional. Two cleavages at about 90°. Triclinic. Often difficult or impossible to distinguish one metamorphic feldspar from another, as twinning and clear grain-shapes are rare.	
K-feldspar	Sometimes yellow—pink (HG) or green—brown (VHG).	**CDE**G
Albite	Can develop clean, angular porphyroblasts (white or a plastic-like blue—grey) even at LG.	**B**CDEG
Plagioclase	Sometimes as inclusion-filled ovoid porphyroblasts in schists (as cordierite). Igneous feldspars are common remnants, and may show distinctive shapes and twins. All feldspars may be reddened, e.g. near faults.	A**BCG**
Pyroxenes Clino (Monoclinic):	Two equivalent cleavages, at 90°, along grain length.	
Diopside	White, grey or pale green. Ovoid, or stubby prisms. (Igneous relics, augitic or diopsidic, are common in meta-gabbros.) Aggregates glassy.	A**BFG**
Jadeite	Grey to pale jade green, as aggregates or needles. Fibrous or glassy.	**H**
Omphacite Aegirine-Jadeite	Deep green. As aggregates or long prisms. Transitional to jadeite.	**H**
Hedenbergite	Dark brown or dark green. In iron-rich rocks and skarns.	
Ortho (Orthorhombic): Enstatite	Grey, or green. (Igneous bronzite: bronze.)	**A**
Hypersthene	Brown.	**B**CE

Table 8.3 (*Continued*)

Amphiboles (5-6)	Two equivalent cleavages, at 125°, along grain length.	
Clino:		
Hornblende Cummingtonite	Deep green, brown or black. Short blades or squat prisms.	**AB**CG
Tremolite	Grey or green blades or needles. Sometimes feathery or fibrous.	AFG
Actinolite	Deep green or green—black. Shapes as tremolite. Asbestos will powder if crushed.	**BCFG**
Blue-amphiboles	Glaucophane is blue. Others are blue—purple, violet, or purply black. Shapes as tremolite. Colour can be partially masked by rims altered to green amphibole. Blue asbestos: DO NOT ATTEMPT TO POWDER—Dangerous if inhaled.	**H**
Ortho:		
Anthophyllite	Grey or pale green. Fibres, needles or blades, often radiating.	**A**
Epidote Group (6)	Prismatic, with one cleavage along length.	
Epidote	Straw to epidote green. } Parallelogram sections, with cleavage along one side. Monoclinic	**BCG**
Clinozoisite	White, grey or pale straw. }	B**G**
Zoisite	White, grey or pale green. Symmetrical cross-sections. Orthorhombic.	B**G**
Olivine (7)	Pale olivine green. Quartz-like, but with a tendency to alteration, to serpentine. Orthorhombic.	AF
Serpentine minerals (2-3)	(Often contains embedded small grains of harder minerals.) Pale to dark serpentine green. Massive or fibrous. Asbestos will mat if crushed.	AF
Chlorite (1-3)	Chlorite green. One cleavage. Cleavage flakes bend, not elastic.	ABCEFG
Talc (1)	White or pale greenish. Feels greasy. One cleavage. Cleavage flakes bend, but not elastic.	AF
Micas (2.5)	One cleavage. Cleavage flakes elastic (springy). Pseudohexagonal.	
Biotite	Green—black or brown—black.	BCDEG
Phlogopite	Yellow, brown or green. In marble or ultramafic rock.	AF
White micas	White, grey or various pale colours (often greenish). Phengite variety. Green, occasionally brown.	CDEFG F
'Sericite' (1-3)	Fine-grained, white or green. Mica-like or talc-like minerals (including some white micas and phengites).	CDEFG
Gypsum (2)	White or grey. One cleavage. Often fibrous. Simple twins.	F

8.2.2 'Metamorphic minerals', often as grains embedded in bulk minerals (Table 8.4)

Table 8.4 *'Metamorphic minerals', often as grains embedded in bulk minerals.*

Mineral (hardness)	Description	Compositional category (see 8.1)
Garnet group (6-7.5)	Dodecahedral habit. No cleavage. Cubic.	**Bold/highlight common**
'Pyralspite' garnets	Red, brown or purple (common garnets).	A**B**D**EG**H
Grossular	Pink—yellow. Rarely greenish.	G
Hydrogrossular	Pale pink. Often not as identifiable grains.	
Andradite	Yellow, brown or green. In iron-rich rocks and skarns.	AB
ALUMINOUS MINERALS OF SCHISTS:		
Stilpnomelane (3-4)	Deep brown or green—black, with one good cleavage. Biotite-like, but with a poor second cleavage, and brittle cleavage flakes. Thin blades or flakes, but more often in shapeless blobs and patches.	CEH
Chloritoid (6-7)	Very dark green or green—black. One good cleavage. Somewhat biotite-like, but with brittle cleavage flakes. Equidimensional tablet shapes, or flaky aggregates. Pseudohexagonal.	**E+G**
Staurolite (7)	Brown prisms, often flat-ended. Twins common. One poor cleavage. Orthorhombic.	**E+**
ALUMINOUS MINERALS OF SCHISTS AND GNEISSES:		
Kyanite (4.5-7)	White or blue glassy blades. Cleavages along and across the length, at 85°. Simple parallel twins. Triclinic.	**DE E+**
Andalusite (7-7.5)	Pink, or white or brown, squat or elongate prisms with square cross-sections. Two equivalent cleavages parallel to prism faces. Orthorhombic. Alters to off-white easily. Sometimes with inclusions in the shape of a cross ('chiastolite').	CD**E E+**
Sillimanite (6.5-7.5)	Whitish. Fibrous, or in needles or long prisms. Often as fibrous bundles. One cleavage along length, on diagonal of diamond shape of cross-sections. Orthorhombic. (By comparison, zoisite needles are shorter, with more rounded-off ends and corners.)	**CDE E+**
Cordierite (7-7.5)	White or blue, and quartz-like. But tends to alter easily, around edges, in concentric zones, or throughout. In schists, more often as ovoid inclusion-filled porphyroblasts. Pseudohexagonal.	**CE E+**

Table 8.4 (*Continued*)

CALC-ALUMINOUS MINERALS (see also plagioclase and epidote group):		
Zeolite group	White. Massive, fibrous or blocky. Best developed filling veins or cavities, but also replaces plagioclase or albite in otherwise little altered igneous rocks. Belongs to several crystal systems.	**BC**
Prehnite	Prehnite green. Brittle tablets with one good cleavage. Occurrence: as zeolites. Orthorhombic.	**B**
Pumpellyite	Blue—green masses and in veins.	**B**H
Lawsonite	White, pale blue, yellow or pale orange. Equidimensional tablets or short prisms, with square or diamond sections. Two best cleavages (of many) at 90°. Orthorhombic.	**BGH**
MINERALS OF HIGH GRADE MARBLES (see also olivines, pyroxenes, and garnets):		
Wollastonite	White, pale grey or green. Fibrous or splintery blades, with 3 cleavages along length. (Tremolite-like.) Triclinic.	**FG**
Monticellite	Colourless grey. Poor cleavage. Often ovoid. Orthorhombic.	**F**
Melilite	White, grey—green or brown. Poor cleavages. Sometimes square-sectioned prisms. Tetragonal.	**F**
Periclase	Grey—white, yellow, brown or green—black. Spherical or octahedral grains. Cleavage cubes. Easily altered to brucite. Cubic.	**AF**
Brucite	White or greenish—grey. One cleavage. Normally either as streaks and masses in serpentine, or pseudomorphs after periclase in marble. (Less soft than talc, and ever found in contact with quartz.)	**AF**
VERY HARD MINERALS:		
Sapphirine (7.5)	Light blue or green. Poor cleavage. Monoclinic.	**ACE**
Spinel (8)	Black or red. Octahedral grains. No cleavage. Cubic.	**A**FG
Topaz (8)	Yellow or white prisms. One cleavage, at 90 to grain length. Orthorhombic. Mainly metasomatic, with fluorite and micas.	
Corundum (9)	White or yellowish, or various colours. Twins. Barrel-shaped grains. Trigonal. In xenoliths and at very hot igneous contacts.	
OTHER METAMORPHIC MINERALS:		
Scapolite (5-6)	White, or various pale tints. Often cloudily altered. Squat or elongate square-sectioned prisms. Tetragonal. Can be indistinguishable from andalusite, but occurs only in rocks with some calcic component, where andalusite cannot.	**BG**

Table 8.4 (*Continued*)

Idocrase (6-7)	Brown or green to yellow. Squat Square-sectioned prisms, or as compact masses. Tetragonal.	**AG**
Humite group (6)	Deep brown, orange or yellow. Chunks; or as fibres amongst fibrous diopside or tremolite. Cleavages poor.	**AF**

8.2.3 'Accessory minerals', bearing minor chemical components (Table 8.5)

Table 8.5 'Accessory minerals', bearing minor chemical components.

Mineral (hardness)	Description	Compositional category (see 8.1)
Tourmaline (7)	Black striated prisms, needles and fans. More glossy than amphiboles. Poor cleavage. Trigonal.	CDE
Apatite (5)	Off-white. May cleave. Hexagonal. Ubiquitous.	All.
Sphene (5)	Grey—brown wedge-shaped grains, or shapeless. Two cleavages. Simple twins. Monoclinic.	ABCDE
Rutile (6)	Red—brown to black. Tetragonal. Pale to red-brown streak.	CDEH
Hematite (5-6)	Red to black. Thin splinters are blood-red. Trigonal. Red—brown streak.	All.
Ilmenite (6)	Black. Igneous relic. Alters to sphene and magnetite. No cleavage. Trigonal. Black or brown streak. Weakly magnetic.	ABC
Chromite (5.5)	Brown—black. Igneous relic. Cleaves. Cubic. Dark brown to black streak. Not magnetic.	A
Magnetite (6)	Black. No cleavage. Octahedral grains. Cubic. Black streak. Strongly magnetic.	All.
Pyrite (6)	Brass yellow. Striated cubes, 'pyritohedra', or rarely octahedra. Cubic. Greenish, grey—black or brown—black streak. (Insoluble in hydrochloric acid.)	All.
Chalcopyrite (4)	Brass yellow, deeper than coexisting pyrite. Shapeless. Very rarely as cubes. Pseudocubic. Green—black streak.	All.
Pyrrhotite (4)	Brownish bronze. Pseudohexagonal. Black streak. Magnetic. Soluble in hydrochloric acid (giving hydrogen sulphide).	ABE
Graphite (1-2)	Grey—black. Feels greasy. Will mark paper with grey—black streak.	EFG

8.2.4 Summary of main rock types and minerals (Table 8.6)

Table 8.6 *Summary of main rock types and minerals.*

Rock-type Name	Main Mineral	Some Likely Minerals	Category (see 8.1)
Amphibolite	Hornblende + plagioclase	Garnet, biotite, clinopyroxene, quartz, clinozoisite	**B**
Anorthosite	Plagioclase (calcic)	Garnet, pyroxenes, amphiboles	**B**
Blueschist	Any assemblage including blue amphibole	Chlorite, epidote, albite, phengite, carbonates, garnet, aegirine—jadeite, lawsonite	**H**
Eclogite	Omphacite + garnet	Quartz, amphiboles, carbonates, clinozoisite, micas, kyanite, olivine	**H**
Granulite	Any assemblage including ortho- and clinopyroxene		
	Acid granulite:	Quartz, K-feldspar, amphibole	**CE**
	Basic granulite:	Plagioclase, olivine, garnet, amphibole, spinel	**AB**
Greenschist	Albite + epidote + either actinolite, or chlorite, or both	Quartz, biotite, garnet, carbonates, white mica	**B**
Quartzite	Quartz	Micas, feldspars, garnet, Al-silicates	**D**
Marble	Carbonate minerals Calcite marble: calcite Dolomite marble: dolomite	Numerous possibilities. Names of carbonates are not additional minerals. See Section 8.1, **F** and **G**	**FG**
Micaschist	Micas (often with quartz or carbonate or both)	Albite, K-feldspar, chlorite, garnet, aluminous minerals, actinolite, graphite	**CDEG**
Peridotite	Olivine	Pyroxenes, amphiboles, garnet, spinel, plagioclase, serpentine, talc, chromite, magnetite	**A**
Serpentine or Serpentinite	Serpentine minerals	Talc, carbonates, actinolite, diopside, chlorite, quartz, olivine, brucite	**A**

8.2.5 Checklist for recording textures and fabrics

1. Keywords to introduce a general impression of the rock:
   Any applicable *fabric or structure-dependent name* (e.g. hornfels, slate, schist, gneiss):
   Whether apparently *isotropic* or *anisotropic*:
   Any *compositional patchiness* (e.g. augen, ribbons, stripes, fine-banding):
   Any *fissility* (schistosity; or cleavage, slaty, spaced, fracture, pencil, etc.):
   Fabric type (slate, mineral, shape, lamination, crenulation, pressure solution stripe) and *symmetry*:
2. *Detailed sketches* of examples of textures and elements of structure if visible in the field.
3. *Statements of time relationships* that are evident in the (sketched) textures and structural elements.
4. *Statements of the grain texture or fine structure features to which fabrics correspond*, with mineral names where applicable.
5. *Any further characterisation of fabrics* (e.g. strings of grains on a schistosity plane; or a stretching lineation on the cleavage of a slate).
6. *Orientations* of all composition and fabric planes, linear features, and intersection directions.
7. *Structural relationships* between fabrics, compositional bands and patches, and deformation bands. Shear sense of shear zones. Refraction. Ages.
8. *Sense of asymmetry* of all intersections of fabrics, other fabrics and banding, and the displacement sense of crenulations, etc., not yet noted.
9. *Correspondence of local fabrics* (their minerals, directions, and structural relations) *to regional patterns* elucidated by mapping and structural synthesis.

8.3 Further Mapping Advice; Formations, Markers, and a Final Report

A good geological description does not demand that equal time be spent on equal geographical area, or equal area of rock exposure. It may not even result from concentrating on those rocks with the greatest information content. It may be necessary to spend time searching for important clues in rocks where information is scarce. Try to work out as early as possible which localities are likely to be worth spending time on, particularly for seeking out and displaying key relationships for regional synthesis and correlation. These may include the following.

1. Geometrical relationships such as which units are exposed along lithological contacts, or the direction of a metamorphic fabric in the hinge of a fold.
2. Evidence of the nature of region- or local-scale variation in rock composition or metamorphic grade.
3. Evidence of the nature of potential pre-metamorphic unconformities or igneous contacts, and the existence of relict sedimentary structures such as particular bands or beds.
4. Evidence of either previous metamorphic (or earlier) minerals, or previous fabrics, possibly occurring for example in relatively unmetamorphosed or undeformed pods.
5. Changes in mineralogy and mineral abundances between two rock formations.

If you have the luxury of sufficient time to study a sub-area repeatedly, then you might want to think about separation by timetable of three roughly equal stages for each sub-area: (i) general mapping and recording obvious lithological or structural features: (ii) concentrated study of complicated exposures, banded sequences and contacts noted during initial mapping, with the aim of developing a deeper understanding of the area of interest: and (iii) construction of a very detailed description, including detailed sketches, maps, and logs of chosen key localities that present prime examples of relationships discovered, primarily for the purpose of compiling a final report.

Any of these phases might involve sample collection, which may be an important objective if you want to subsequently undertake laboratory-based research on important samples from your field area (for instance trying to ascertain the age of metamorphism, the pressure and temperature conditions of metamorphism, or the extent of fluid interaction with rocks in the region). Note that you may need permits for sampling in some areas, and these should be obtained well ahead of the planned fieldwork. Above all, keep in mind the overarching purpose of the fieldwork and any specific questions that you are trying to use it to address. This will help you to hone in on the key observations that may make the fieldwork successful.

So, in summary, when planning a timetable, you need to make sure that you allow sufficient to time to:

1. Cover the field area and map it, divided into convenient sub-areas.
2. Record the particular exposures showing a lot of detail.
3. Record and display the characteristics of potential key localities.
4. Work out the overall character of banded sequences.
5. Inspect closely all the contacts, boundaries, and faults, to record any spatially related features.
6. Collect any samples that may be important later.
7. Have 'injury time' in reserve, for unforeseen difficulties, geological or otherwise.

The timetabling of these tasks will depend on the total time available, the size of the field area, the degree of exposure, the amount of 'down time' likely to be required because of inclement weather, and numerous other factors that cannot always be easily controlled.

8.3.1 Formations

Production of good maps and field sections is essential to the description of any field area. This involves, fundamentally, dividing the rock mass into 'formations' (mappable units) with precisely specified diagnostic features, and displaying their geometrical pattern on maps and sections. In the case of sequences that have experienced metamorphism, there are several additional challenges, such as deciphering how the current disposition of a sequence of rocks relates to inherited protolith chemistry and structure, understanding the effects of metamorphic overprinting, and recognizing the loss or addition of fluid or melt. Furthermore, you may need to determine why rocks change throughout an area: is it because the intensity of metamorphism is varying, because the protolith lithology is different, because only one part of the sequence had melt extracted, or are you seeing a combination of all of these effects?

As such, formation definitions and distinctions are somewhat more complicated for metamorphic rocks than for sedimentary or igneous ones. It is not easy to fit metamorphic rocks into a simple stratigraphic time line like that of sedimentary sequences. The final rock clearly had both pre-metamorphic and metamorphic histories, possibly as well as a deformational history. There may even have been multiple overprinting successions of deformation and/or metamorphism. It is thus necessary to establish a clear hierarchy of distinctions whenever more than one kind of criterion is used to characterize formations. For example, a distinction may be made at group level between a meta-sedimentary group and a meta-igneous group. At formation level, the meta-sediments may be divided by metamorphic grade, and the meta-igneous rocks by grade and possibly by division between 'massive', 'banded', and 'banded-and-strongly-veined' formations. The reality will vary considerably depending on the style of metamorphism being studied and the complexities of the outcrop. It is important to take into account any lateral and vertical variations within each unit before assigning them to a formation.

Possible kinds of criteria for distinguishing formations include:

1. Pre-metamorphic rock-types (the protolith), including preserved igneous and sedimentary structures.
2. Metamorphic grade, either by the presence or absence of index minerals or by change of the entire mineral assemblages (see Chapter 3 for details).
3. Existence or intensity of deformation fabrics, or differences of structural style (e.g. brittle or ductile). Evidence of shear-sense indicators (see Chapter 5 for details).
4. Grain size, either as an average grain size for the rock or in terms of one or several minerals occurring as porphyroblasts (see Chapters 3 and 5 for details).
5. Degree of homogeneity or heterogeneity: is the rock banded, massive, or patchy? Can you see obvious alignment of mineral grains, separation of a gneissose fabric, or melt segregations (see Chapter 4 for details).
6. Colour (it can be useful to note both weathered and fresh surfaces, but fresh surface colour should be used).
7. Style of weathering or fracture.
8. Presence or absence of veins, pegmatites, dykes, boudins or pods.

The word 'Formation' makes it clear that you are referring to a rock unit, not a rock type. You should always identify and decide on your own formation contacts and names in the field, based on your observations. You will need to define marker horizons (described further in Section 8.3.2) on your own terms and these should be explained in your notebook and in any final written report. Once you have mapped formations in your field area, you can compare them to previous work to see where similarities and differences occur. You will usually find that many of your formations are in agreement with previously mapped ones, but difficult formation contacts (e.g. gradational ones) and different scales of mapping may highlight some different formation definitions. Formation names should consist of either two or three parts: Name and Rock type (e.g. Vishnu Schist), or Name, Rock type, Formation (e.g. Gleasbrook Quartzite Formation). More on rock type information can be found in Chapter 3.

8.3.2 Markers and stratified sequences

The geometric pattern of rocks, at a suitable scale for representation on the map, can generally be constructed by tracing out 'markers' such as unusual or distinctive lithological layers. Such marker bands may be sufficiently thick that their boundaries can be accurately represented on the map. Otherwise, they must be marked conventionally by a dashed line (see common mapping symbols in Figure 2.9). The marker may be a sudden change in rock type, the plane separating two rock types being a marker horizon. Or it may be the first occurrence of an indicator mineral or melt, as in the case of Barrow's classic (1912) work (see Figure 1.3). For practical purposes, any gradational change sharp enough to take place over a distance too small to be representable on the map, or any very thin marker band, can also be used as a marker horizon (as can sharp boundaries). When the boundary is gradual over some distance you can specify a point along that transition to define a marker (e.g. 50 : 50 ratio between the gradational rock types). Markers should be followed in the field and drawn directly onto your field maps whenever possible. In most cases the markers will also coincide with your formation contacts, but not always (e.g. markers may delineate groups/members within your formations).

The concept of markers derives from, but is not restricted to, originally stratified rocks. If layered rocks exist in the area, it is likely that some division of the sequence into formations, and identification of marker bands, may be needed. This should be done with full awareness of potential structural complications (e.g. thrust repetitions or inversions on folds). Two similar bands may well turn out to be parts of the same original layer. Any unusual but recurring rock type should be noted early on. If layered sequences occur only at two widely separated outcrops, there is likely to be a problem in deciding whether they are the same formation or not. The same pattern of markers may well be the evidence that correlates them, and this possibility should be searched for. Possible contrary evidence, such as non-equivalence of markers, or large differences in metamorphic grade, should be noted at the same time. In general, it is good practice to tackle banded sequences early on, as they can show features and highlight potential difficulties that will help in later handling of outcrops of less informative rocks. Also, where possible, you may find it easier to map less deformed or less structurally complex formations first.

8.3.3 Developing the big picture of the area

At the later stages of fieldwork, considerable thought should be given to the regional-scale geological framework within which the rocks will be reported. With a nearly completed field map and sections, good descriptions of formations and contacts, and with representative logs and sketches, it should be possible:

1. To propose a geological history of the rock mass.
2. To devise a hierarchy of rock units which reflects both real geological affinities and easy organization of material.

Make sure that you consider potential alternative histories and hierarchies in order to understand those geological relationships that remain ambiguous. Search for any discrepancies between individual field records and the generally implied historical and geometrical relationships between units. Remember that *a complete field record of an outcrop will be such that additional visits to that outcrop will probably not be required unless a dramatically new interpretation cannot be reconciled with observations.* The most vital individual observations that refute an otherwise acceptable geological synthesis may nevertheless deserve to be confirmed at the end of fieldwork, and perhaps recorded in greater detail in order to convince readers of your interpretation.

8.3.4 Remember to use your detailed observations

It can be relatively easy for a final report to fall into a fairly sterile description of the geology and review of previous work published in the area. One of the clear advantages of undertaking detailed mapping and fieldwork is that you will be often making observations and taking measurements that have never been done or that provide a new view of a pre-existing idea, that all may go to change opinion and move the science forward. To this extent, is important that you use the data you collected in the field as much as possible within your write up.

Do not be afraid to use details such as sketches that you have made as figures within your report. A well-drawn sketch with the notebook can act as a great template for a valuable figure in your report, which also helps to marry the two records of the geology (e.g. Figure 8.1). Other data such as structural measurements can be neatly plotted (e.g. on stereonets) for additional graphical representation of your measurements (e.g. Figure 8.2). This can also enable you to compare your work with published examples where appropriate. Another good hint when writing up, particularly in this digital age, is to make good use of your field photographs. Additional sketches and interpretations can be drawn up as overlays within a graphics package, and with clever use of sketching you can quickly add significant value to your photos and also help to show how multiple scales of photograph relate to each other (e.g. Figure 8.3).

There are always a number of different ways to represent the data you have collected in your notes, maps, and photo collections within your report writing. A suggested layout for a report is presented in the next section, but we note that you can learn a lot by spending some time flicking through some other examples of reports/theses, etc., to gain some ideas about structure and presentation. Remember to pick only 'good' examples to learn from, and hopefully some the best practice ideas will help you develop your own skills. Good Luck!!

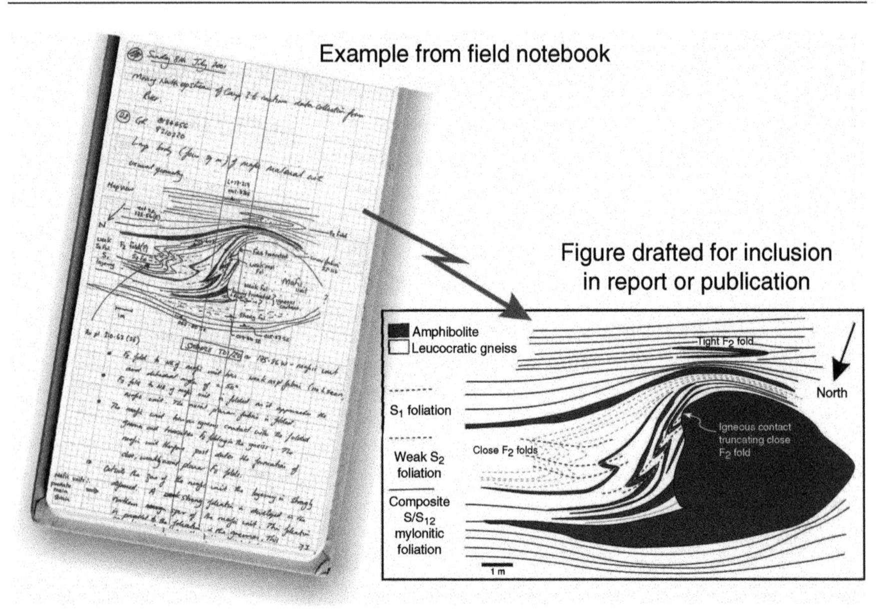

Figure 8.1 *Examples of a detailed field sketch being neatly drawn up in a graphics package to be used in report/publication (example courtesy of Steve Reddy).*

Final report: suggested layout

1. *Introductory statements:*
   The area studied, and its boundaries.
   Who did the work, and when. The base maps used, their scale, source, and publication dates. Any relevant peculiarity of the work conditions (e.g. abnormal snow cover; the use of climbing equipment; forms of transport).
2. A brief statement of the *geographical layout* of the outcrops of different units, their relationships to topography, degrees of exposure, general weathering condition, etc.
3. Statements, preferably in the form of a table, of the *hierarchy of units* to be used for detailed descriptions, of the *definitions of formations* found on the fair-copy maps and sections, and of the *correspondence between mapped and described units*. Where stratified rock units are concerned, an overall stratigraphic column, drawn to scale, is a convenient way of displaying the relationship between them.
4. *Descriptions of rock units*. Characters common to a group of units may be described first (e.g. common metamorphic grade or fabric orientation), followed by the description of units in order, subdividing as necessary. If in doubt, use the order of features in this book (banding, minerals, texture, etc.).
   Each description may be based on a detailed example map or a detailed log (chosen either as typical or illustrating extremes), together with statements of variations.
   Take care to state clearly how characters of the rock change on approaching and reaching each contact.
5. *Synthetic statements* of geometrical patterns and distributions (e.g. stratigraphic correlations made on the basis of field evidence; distributions of sedimentary facies; patterns of igneous intrusions; pattern of metamorphic grade; structural synthesis; distribution of mineralisation).

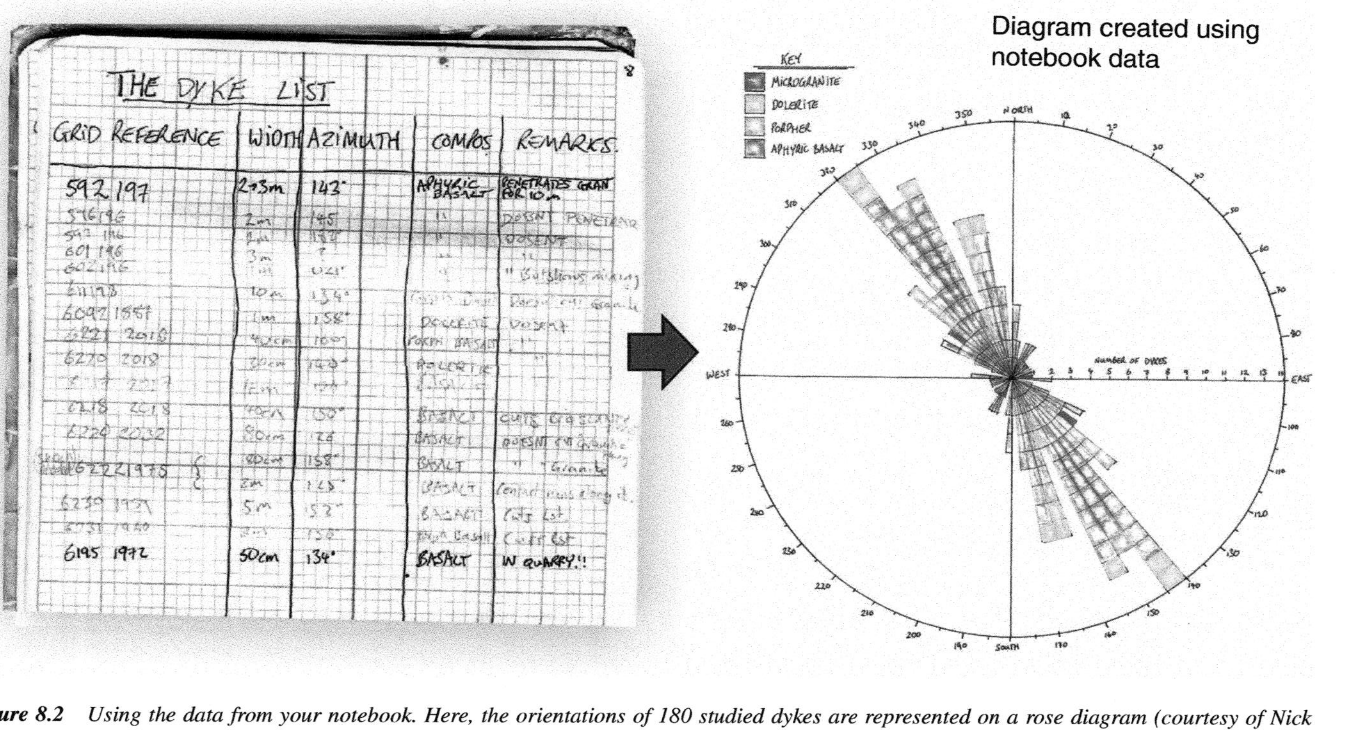

Figure 8.2 *Using the data from your notebook. Here, the orientations of 180 studied dykes are represented on a rose diagram (courtesy of Nick Timms). Note that data have been repeated with 180° of rotation for symmetry.*

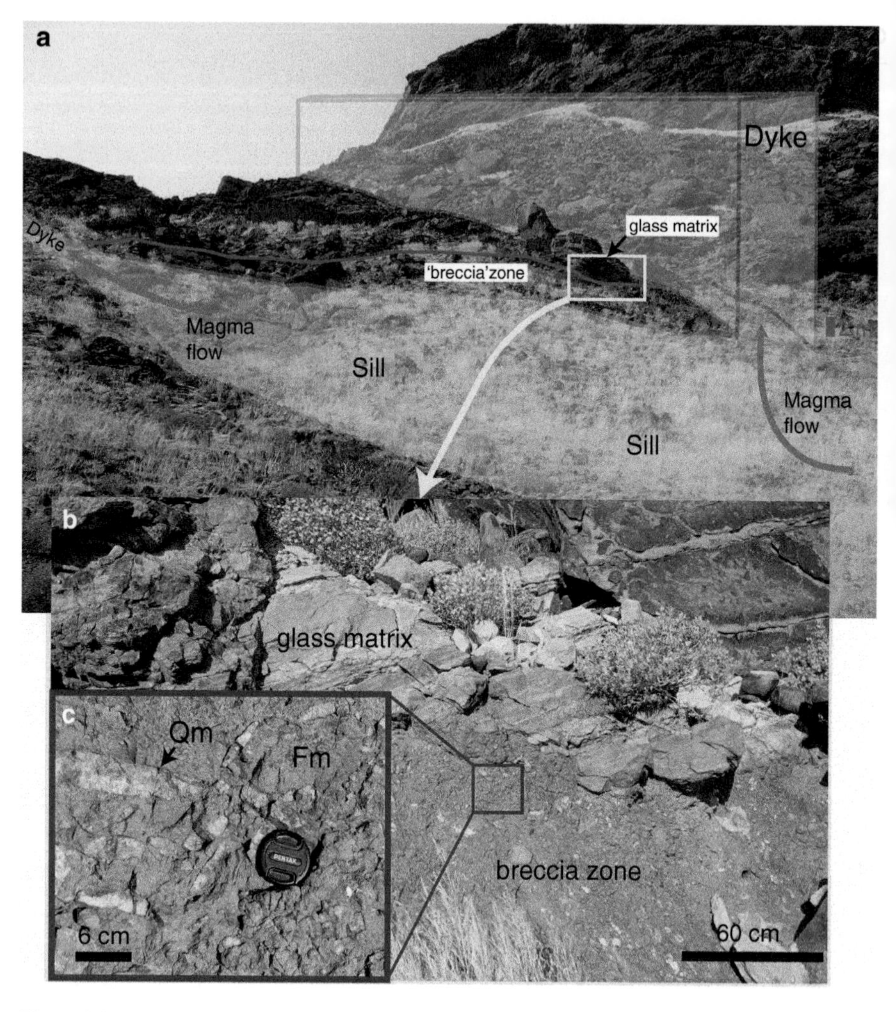

Figure 8.3 *Using overlays/sketches and linked photos in a figure for a report/thesis. This example is of intense contact metamorphism in NW Namibia and is organized as a composite of three images spanning scales from decametres to centimetres. (a) A broad overview image facing east towards the outcrop. Igneous intrusions have been highlighted in green and the dyke geometry has been extrapolated in 3D. (b) Closeup image of contact between segregated breccia-like melt and glass matrix zone. Scale bar is 60 cm. (c) Overview of the breccia-like melt segregation with a super closeup on the melt segregation 'breccia' (Qm = quartz matrix, Fm = feldspar matrix) (courtesy of Clayton Grove).*

6. *Geological history*, stating *what has and what has not been established.* Possible points of correlation of this history to that of other areas, or other studies.
7. *Fair-copy maps and cross-sections.*
8. *All field documents*, in the state they left the field (notebooks, field maps and sections, detailed plans and logs, etc.). These should normally be accompanied by representative *rock samples*, listed and referred to in the descriptive text.

Checklist of Rock Features

ASSOCIATION:	Chapter/Section
Major rock forms (mapping).	2
Gross compositional banding.	4.1
Further sedimentary or igneous features (chilled margins, xenoliths, cobbles, cross-bedding, grading, etc.).	–
Fine banding or striping.	4.2
Pre-metamorphic association name.	3.3, 8
INDIVIDUAL ROCK TYPE:	
General properties: Colour, jointing, fracture, schistosity, weathering style and staining, hardness, magnetic.	–
If minerals visible: Minerals recognized.	3.1, 8.2
Which occur adjacent, and which do not.	
Grain size of each mineral. Porphyroblasts.	5.2
Proportions of minerals, variability, patchiness, aggregates.	
Fine-grained material.	3.2
Rock type name.	3.3, 7.2, 8.3
Details: peculiarities of minerals, details of unknowns.	3.4
Texture, anisotropy and time relations between minerals.	5.2, 8.4
DEFORMED ROCKS:	
Texture, fabric, and fissility	4.2, 5.3, 8.4
Orientations, relations of fabrics and structural elements.	4.4, 5.5, 8.4
Scattered components of deformed rocks:	
Shear pods, boudins, undeformed masses.	5.1
Large grains, augen, flaser, grain pressure shadows.	5.2
SCATTERED ROCK COMPONENTS (General):	
Pseudomorphs.	5.3
Veins and minor intrusions. Geometry and internal features.	6.3
Igneous contacts. Aureoles and metasomatism.	6.1
Contacts, reaction, and reaction zones.	6.4
FAULTS, FAULT ZONES, AND FAULT ROCKS:	
Fault relations.	7
Fault materials in general.	7.1
Fault rocks.	7.2

FURTHER READING SUGGESTIONS

General helpful texts (including books that have helped inspire us with this revised edition):

Best, M.G. 2002. *Igneous and Metamorphic Petrology*, 2nd Edition. Wiley-Blackwell. 752 Pages. ISBN: 978-1-405-10588-0.

Coe, A.L. (Ed.) 2011. *Geological Field Techniques*. Wiley-Blackwell. 336 Pages. ISBN: 978-1-444-34823-1.

Hefferan, K., & O'Brien, J. 2010. *Earth Materials*. Wiley-Blackwell. 672 Pages. ISBN: 978-1-444-39121-3 (note: new edition due out in 2022).

Jerram, D., & Petford, N. 2011. *The Field Description of Igneous Rocks*, 2nd Edition, Wiley-Blackwell. 256 Pages. ISBN: 978-0-470-02236-8.

Lisle, R.J., Brabham, P., & Barnes, J.W. 2011. *Basic Geological Mapping*, 5th Edition. Wiley-Blackwell. 232 Pages. ISBN: 978-1-119-97751-3.

Locatelli, M., Verlaguet, A., Agard, P., Federico, L., & Angiboust, S. (2018). Intermediate-depth brecciation along the subduction plate interface (Monviso eclogite, W. Alps). *Lithos*, 320–321, 378–402.

Passchier, C.W., & Trouw, R.A.J. 2014. *Microtectonics* (2nd rev. and enlarg. ed. 2005, repr. ed.). Berlin: Springer.

Philpotts, A.R., & Ague, J.J. 2009. *Principles of Igneous and Metamorphic Petrology*, (2nd ed. ed.). Cambridge, UK: Cambridge University Press.

Sawyer, E.W. 2008. *Atlas of Migmatites*. Ottawa, Ontario: NRC Research Press & the Mineralogical Association of Canada.

Spear, F.S. 1995. *Metamorphic Phase Equilibria and Pressure-Temperature-Time Paths* (2nd print, corr. ed.). Washington, D.C.: Mineralogical Society of America.

Winter, J.D. 2010. *Principles of Igneous and Metamorphic Petrology* (2nd ed. ed.). New York: Prentice Hall.

Yardley, B.W.D. 1989. *An Introduction to Metamorphic Petrology*. Harlow, Essex, England: Longman Scientific & Technical.

Advanced reading –good research papers and reviews that cover both field & laboratory-based studies integrated with theoretical frameworks

Brown, M., Korhonen, F.J., & Siddoway, C.S. (2011). Organizing melt flow through the crust. *Elements*, 7(4), 261–266. doi:10.2113/gselements.7.4.261

Brown, M. (2007). Crustal melting and melt extraction, ascent and emplacement in orogens: mechanisms and consequences. *Journal of the Geological Society*, 164(4), 709. doi:10.1144/0016-76492006-171

Carlson, W.D., Pattison, D.R.M., & Caddick, M.J. (2015). Beyond the equilibrium paradigm: How consideration of kinetics enhances metamorphic interpretation. *American Mineralogist*, 100(8–9), 1659–1667.

Cesare, B. (1999). Multi-stage pseudomorphic replacement of garnet during polymetamorphism: 1. Microstructures and their interpretation. *Journal of Metamorphic Geology*, 17, 723–734.

The Field Description of Metamorphic Rocks, Second Edition. Dougal Jerram and Mark Caddick.

Clark, C., Fitzsimons, I.C.W., Healy, D., & Harley, S.L. (2011). How does the continental crust get really hot? *Elements*, 7(4), 235–240. doi:10.2113/gselements.7.4.235

Ferry, J. M., & Spear, F. S. (1978). Experimental calibration of the partitioning of Fe and Mg between biotite and garnet. *Contributions to Mineralogy and Petrology*, 66, 113–117.

Holder, R.M., Viete, D.R., Brown, M., & Johnson, T.E. (2019). Metamorphism and the evolution of plate tectonics. *Nature*, 572(7769), 378–381.

Kohn, M.J. (2016). Metamorphic chronology—a tool for all ages: Past achievements and future prospects. *American Mineralogist*, 101(1), 25–42.

Locatelli, M., Verlaguet, A., Agard, P., Federico, L., & Angiboust, S. (2018). Intermediate-depth brecciation along the subduction plate interface (Monviso eclogite, W. Alps). *Lithos*, 320–321, 378–402.

Powell, R., & Holland, T.J.B. (2008). On thermobarometry. *Journal of Metamorphic Geology*, 26, 155–179.

Putnis, A., Moore, J., Prent, A. M., Beinlich, A., & Austrheim, H. (2021). Preservation of granulite in a partially eclogitized terrane: Metastable phenomena or local pressure variations? *Lithos*, 400–401. doi:10.1016/j.lithos.2021.106413

Reddy, S.M., Collins, A.S., Mruma, A. (2003). Complex high-strain deformation in the Usagaran Orogen, Tanzania: structural setting of Palaeoproterozoic eclogites. *Tectonophysics*, 375, 101–123.

Sawyer, E.W. (2001). Melt segregation in the continental crust: distribution and movement of melt in anatectic rocks. *Journal of Metamorphic Geology*, 19, 291–309.

Sawyer, E.W., Cesare, B., & Brown, M. (2011). When the continental crust melts. *Elements*, 7(4) 229–234. doi:10.2113/gselements.7.4.229

Spear, F.S., Pattison, D.R.M., & Cheney, J.T. (2016). The metamorphosis of metamorphic petrology. In M.E. Bickford (Ed.), *The Web of Geological Sciences: Advances, Impacts, and Interactions II*. Geological Society of America Special Paper 523.

Waters, D.J., & Lovegrove, D.P. (2002). Assessing the extent of disequilibrium and overstepping of prograde metamorphic reactions in metapelites from the Bushveld Complex aureole, South Africa. *Journal of Metamorphic Geology*, 20, 135–149.

Wheeler, J., Mangan, L.S., & Prior D.J., 2004. Disequilibrium in the Ross of Mull Contact Metamorphic Aureole, Scotland: a Consequence of Polymetamorphism. *Journal of Petrology*, 45(4), 835–853.

White, R.W., & Powell, R. (2002). Melt loss and the preservation of granulite facies mineral assemblages. *Journal of Metamorphic Geology*, 20, 621–632.

White, R.W., & Powell, R. (2011). On the interpretation of retrograde reaction textures in granulite facies rocks. *Journal of Metamorphic Geology*, 29(1), 131–149.

Whitney, D. L., Teyssier, C., Rey, P., & Buck, W. R. (2012). Continental and oceanic core complexes. *Geological Society of America Bulletin*, 125(3-4), 273–298. doi:10.1130/b30754.1

Yardley, B. W., & Bodnar, R. J. (2014). Fluids in the continental crust. *Geochemical Perspectives*, 3(1), 1–2.

Wiley Blackwell also hosts the Journal of Metamorphic Geology, *showcasing some of the recent advances in the subject. Edited By: Michael Brown, Katy Evans, Doug Robinson, Richard White, and Donna Whitney. Print ISSN: 0263-4929 Online ISSN: 1525-1314.*

INDEX

The Field Description of Metamorphic Rocks, Second Edition. Dougal Jerram and Mark Caddick.